Y0-CWI-810

Bluebeam Revu eXtreme 20 Lab Manual

prepared by

Gregory D. Kelly

JohnWyatt Publishing

gregory.kelly@cincinnatistate.edu

Preface

This manual is intended to be used as a lab guide for Bluebeam Revu eXtreme 20 or be used as a stand-alone tutorial guide for using Bluebeam Revu eXtreme 20. It was written for Bluebeam Revu eXtreme 20 but can be used, with *very slight adjustments*, with other versions of Bluebeam Revu. Prior to 2018, Bluebeam Revu had a slightly different interface, but much of the same functionality. With some minor translation, one can easily apply much of this manual to older versions.

This manual assumes the use of Bluebeam Revu eXtreme edition. Bluebeam Revu also in the Standard and CAD editions. If you do not have Bluebeam Revu eXtreme, you may find some functionality to be different.

Detailed below are the various conventions adopted for use in the manual:
- Throughout this manual, the Bluebeam Revu eXtreme 20 software will be referred to simply as Bluebeam (with a capital "B") or as "the software"
- Data that is shown in this **Arial, bold** font is intended to be manually entered by the user throughout the various exercises of the manual
- Content (button, tool, etc.) that is meant to be selected on the Bluebeam Revu interface will be referred to in unbold Arial, explicitly as it appears in the software.

Table of Contents

Lab 1

The Interface

Getting Started Using

In order to perform the exercises in this lab manual, a file of sample documents has been provided. These files will need to be accessible while using Bluebeam. Take the time now to download and save the files in a location that you will remember and that is easy to find.

Opening the Software

Follow these steps to open Bluebeam Revu eXtreme 20.

1. Bluebeam Revu eXtreme 20 can be accessed by simply opening a PDF document you wish to use in Bluebeam. To do this, navigate to where you have saved the document you wish to use. Right click on the document and select Open with... , and then Revu . If you have set Bluebeam as your default PDF viewer, you can simply open the document.

Note: It is often convenient to open multiple documents at once. Multi-select the files from their location to open them all at once. The documents will open in Bluebeam, each in separate tabs, much like a web browser. We will use this method when we begin soon.

Another option for opening a document in Bluebeam can be to open the Bluebeam application first, without initially opening a document. From your desktop, click on the *Windows* button ⊞ and find Bluebeam Software, then scroll down to Bluebeam Revu eXtreme 20 Bluebeam Revu , or simply type **Bluebeam** into the search field to find it. Once the software is open, select **Open** from the File menu on the Menu Bar. Navigate to where the document is saved, and select the document(s) you wish to open.

A set of files was provided with this manual. These files will be used throughout the manual. We will commonly refer to these files as the "sample files", as they are sample material provided with which to practice the tasks being explained. You will want to save those some place that you can easily access them throughout the manual.

2. Navigate to the sample files, select all, and open all documents included in the Lab 1 folder as shown here:

Note: Notice by the file icon that the default PDF viewer has been set to Bluebeam Revu here.

3. When the software opens, the three files will open each in a separate tab in Bluebeam. The tabs will be found near the top of the window as shown here.

4. With the software open, use the File Access function by clicking the **File Access** button on the left 🖫. Here you can pin frequently used documents, see recently used documents, or navigate to other file locations.

5. With the **File Access** panel open [File Access ∨ / ⏱ Recent Files], multiple sort and filter options are available, as shown here. Click the icon next to [Recents]

Note: Holding the Ctrl key while clicking any document will open that document in the background.

6. Next, pin these three documents by right-clicking on each document and select **Pin,** or simply hover over the file and click the pin icon 📌, and then click [📌 Pin File].

Categories can be used to group pinned documents. When you click the pin icon, select [New Category] and enter the desired category name.

7. As you hover over the first document, click the pin icon 📌, and then select [New Category].

8. In the Category: field type the following: ***Real World Bluebeam Documents.***

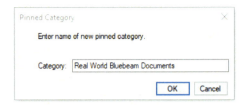

9. Then click OK

10. Next, pin all the rest of the documents provided to this newly created category. Right click on each. Select Pin > and then Real World Bluebeam Documents .

With the File Access panel open, follow the instructions here to perform your first of many screen shots.

Throughout this manual you will see the following:

<u>Perform a screen shot and save to your computer</u>.

When you see this, you will need to take a snapshot of what has been performed to submit. There are many simple ways to complete the task. To keep it simple for all users of this manual, and to ensure consistency of submissions to your instructor, your computers Print Screen function should be used.

<u>Important Note</u>: Your instructor may give you specific instructions for saving and submitting screenshots that differ from the instructions given here. It is strongly recommended that users of this manual clarify the specific process with their instructor before proceeding.

1. When prompted to do so, the Print Screen function should be used by simply pressing the **Print Screen** button on your computer's keyboard.

In order to submit the screen shot to your instructor, you will need to open a blank Word document.

2. Orient the page in landscape as shown here.

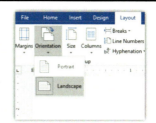

3. Right click and select **Picture** from the Paste Options.

You will need to save those screenshots in a dedicated place on your hard drive preferably with all documents related to this lab manual. Be sure to save them where you can access them in the future to be submitted.

4. For each file or screenshot, use the following naming convention: **Real World Bluebeam Revu – (your name) Lab #1 Screenshot (##)** – numbered sequentially. Therefore, this initial screenshot will be called: **Real World Bluebeam Revu – (your name) Lab #1 Screenshot 1**

Your instructor should provide instructions on how you are to submit your screenshots.

5. With the File Access panel open, perform a screenshot.

* Perform a screen shot and save to your computer.

Your screenshots should look similar to this example.

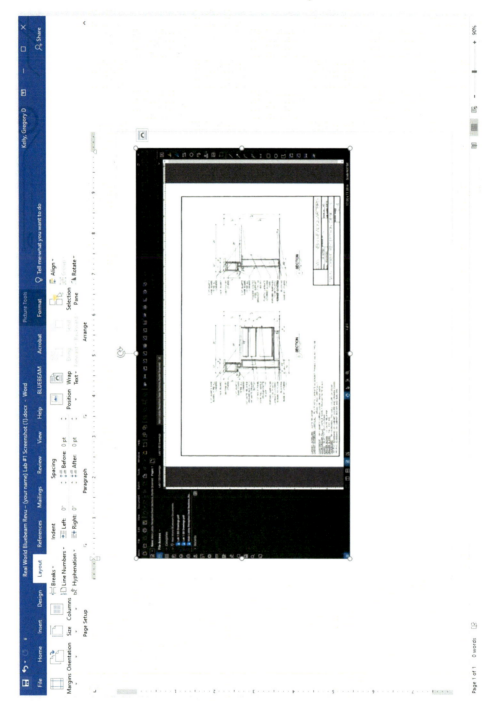

Note: After you finish saving to the hard drive as described in these instructions, it is *also* a good idea to save the file to a flash drive and/or other location as well, ensuring that you have it saved in at least two separate locations.

Vector vs. Raster

It is extremely important to understand the difference between vector data and raster data and how it is used in the documents with which you will view and interact. Bluebeam can use both, but there are important distinctions.

Vector data is made up of lines, curves, points, and paths that are generated by mathematical formulas in digital space. The content can be scaled infinitely.

Raster data is made up of pixels. Each pixel has associated data that is represented in the form of a grid matrix of cells in digital space.

Most documents that were created electronically will be published in vector data unless they have been downgraded or published in a raster format. If a physical document has been scanned, the photographic image of the document will become raster data.

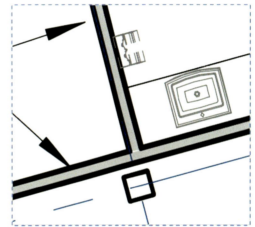

The simplest way to find out whether you're dealing with Raster or Vector data in Bluebeam is to zoom in on the content. If you have a vector document, the content will appear sharp.

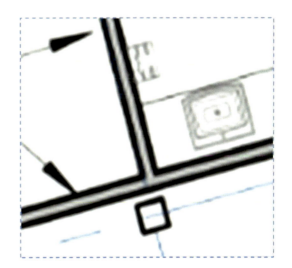

Rasterized data will appear pixelated.

Vector data has some distinct advantages over raster data. Because vector data has mathematical definitions, the data can be referenced in many useful ways, such as snapping to associated content. The content is also referenced in other functions, such as Bookmarks and Hyperlinks. Editing, searching, and selecting PDF content requires vector data.

Some raster data can be salvaged for use with a tool called Optical Character Recognition (OCR) within Bluebeam.

1. From the sample documents, open the file named, Main Lobby Reception Desk Sections_Raster Scanned.pdf

2. Next, select **Document** from the menu bar and select 〔OCR〕

3. In the OCR dialogue box, leave the default selections as they are and click 〔OK〕. Depending on the size of the file, this may take some time.

4. Once complete, use the Select Text 〔IA〕, found on the Navigation Bar at the bottom of the window, select the door/drawer pulls information from the drawings sheet

```
INET INTERIORS = WHITE MELAMINE
INET EXTERIORS (VERTICAL) - PL-I PLASTIC LAMINATE WILSON
                          GRAIN TO RUN VERTICALLY
NTERTOP - QS-I QUARTZ CORIAN, COLOR : SNOW WHITE
R/DRAWER PULLS - HAFELE MODEL# 155.01.611 MATTE BLACK
SES - BLUM 170° SELF CLOSING
WER BOXES - 1/2" WHITE MELAMINE
WER SLIDES - FULL EXT. B.B.
KBOARD - TB-I KOROSEAL TAC-WALL COLOR: 04 STONE
PVC EDGEBAND ON CABINET DOORS AND DRAWER EDGES
```

Then right click to copy and paste into a web search:

〔G HAFELEMODEL#155.01.611MATTEBLACK 〕

5. Now do the same for the hinges.

6. With your web browser and search engine open, perform a screen shot.

* Perform a screen shot and save to your computer.

Interface

There are a few main areas of the Bluebeam interface. See the color-coded screenshot below for the areas referenced:

- 🌀 **Menu Bar**
- 🌀 **Navigation Bar**
- 🌀 **Panel Access Area**
- 🌀 **Toolbars**
- 🌀 **Markups List**
- 🌀 **The Main Workspace**
- 🌀 **Properties**

Menu Bar

In the Menu Bar, features are organized into task oriented groups making almost all tools available through the respective sub-menus.

- The Revu menu contains some important items that will be used to affect the entire program, such as: About, Preferences, Profiles, and Unregister
  Note: Also, in this menu is *Keyboard Shortcuts*. Here you can view and edit the many keyboard shortcuts that can be used to navigate and command throughout Bluebeam.

- The File menu contains commands allowing you to create and manipulate files.

- The Edit menu contains the undo and redo functions, some navigation functions, but is primarily the home of the PDF Content editing commands.

- The View menu contains commands allowing you to manipulate how you view and interact with the document.

- The Document menu contains mainly commands that effect the entire document.

- The Tools menu contains many of the various tools which will often be found on the various Toolbars. This is also where you will go to turn on and off the various Toolbars.

- The Window menu will allow you to control the Main Workspace window.

- The Help menu contains help and other resources.

Navigating

There are several basic navigation techniques that will be useful moving forward. Firstly though, it is recommended that a 3-button mouse, with a scrolling wheel, is used within Bluebeam.

Mouse functionality:

- When opening a PDF, Bluebeam will recognize a word-processed document. The scroll wheel will scroll up and down in the document.

- When opening a large format PDF, the mouse will behave similar to that used in a CAD program. The scroll wheel will zoom in and out. Pressing the scroll wheel will pan.

- You can change the behavior of the scroll wheel temporarily by pressing and holding the Ctrl key while rolling the scroll wheel.

Navigation Bar

Hover over each icon inside of Bluebeam for its name.

From left to right:

- Unsplit: remove split when viewing splitscreen
- Split Vertical & Split Horizontal: Used to split the screen so multiple documents or multiple parts of the same document can be viewed on the same screen in the same Instance of Bluebeam.
- One Full Page: The document will be fit to the screen vertically such that an entire page is viewed.
- Scrolling Pages: The document will be fit to the screen horizontally in width.
- Pan: Left click will pan the document.
- Select: Left click will select or multi-select objects in the document.
- Select Text: Left click will select only text, only if vector text and selectable.
- Zoom Tool: Left click and hold to define extents to zoom.
- First Page: Navigate directly to the first page of the document.
- Previous Page: Navigate to the previous page from the current page.
- Page Field: Enter a specific number to jump directly to that page.
- Next Page: Navigate to the next page from the current page.
- Last Page: Navigate directly to the last page of the document.
- Previous View & Next View: Navigate back and forth between locations and zooms that you have used.
- Page Size: Bluebeam recognizes the physical page size of the document and displays that here.
- Scale: If set, the scale of the page will display here. It may vary or it may not be set.

Panel Access Area

Panels are accessed from icons located down the left side of the interface (shown left). This is the Panel Access Area. Each icon will open its associated panel. The panel width can be adjusted to suit your needs allowing you to maximize screen real estate. You can move a panel by clicking and dragging the panel icon to a new location. We will go into detail on some of these panels as we address their respective features and functions throughout this manual.

Toolbars

Many tools are accessed here. All of the tools available from the Menu Bar are available here as well as others. The Toolbars are fully customizable. They can be moved around to suit your needs. They can be turned on and off as needed. Toolbars can be moved around and customized to suit your preferences.

Markups List

The Markups List is an often-overlooked by extremely powerful feature of Bluebeam. It is easy to miss as it is often in the collapsed state at the bottom of the interface. Hover over the 3 bullet points icon and you'll see it says Markups. Click on this to open the Markups List. We will go into much more detail on this in Lab 3 Tools.

The Main Workspace

The Main Workspace is the primary center for interacting with the document. You can open multiple tabs and open Bluebeam's web browser in this area using the WebTab command under the Window menu.

Properties

Properties is a toolbar located at the top of the Main Workspace and below the Menu Bar that shows you the various properties of the current selection. This is often a quick way to reference the properties and make basic changes. However, be aware that the Properties Panel, located in the Panel Access Area, will have the full array of properties available for any given item.

Split Screen/Split Window

Split Screen and Split Window are two functions that can be extremely useful on occasion.

Split Screen:

Split Screen can be useful if you want to view multiple things on one screen, whether that's multiple documents in different tabs, multiple pages of the same document, multiple views of the same page, or perhaps a document in one tab and the built-in web browser in another tab. A screen can be split horizontally, vertically, or both, up to 16 times. The Split Screen command is probably most conveniently found on the Navigation Bar, but can also be found in the View menu. Under the View menu are other commands to further manipulate the Split Screen, such as: Switch, Balance, or Synchronize. Synchronize can be useful if you want to view the same area of a drawing in multiple documents – perhaps multiple revisions of the same sheet, OR an Architectural plan with the Structural plan of the same area, etc..

Now we'll try using the Split Screen function.

1. Open the project drawings from the sample documents by opening the file named, **Lab 1 DD Drawings.pdf**. The file should open in one full page.

    You can close any other tabs that may still be open at this time. If any panels are open, click on the icon to close the panel.

2. From the View menu, select/check Synchronize Page

3. Select the Split Vertical (⊞) icon from the left side of the Navigation Bar. You should have a vertical split down the middle of your screen. If you hover your curser over the split bar, your curser will become a two-sided arrow curser ↔ . Click and drag to resize your windows.

4. Click on the tab at the top of the left window, then navigate to page 8 of the drawings, which should be A101 FLOOR PLAN – FIRST FLOOR.

5. Next, click on the tab at the top of the right window, then navigate to page 11 of the drawings, which should be A121 REFLECTED CEILING PLAN – FIRST FLOOR.

6. Click in the left window and zoom and pan into the area containing the restrooms:

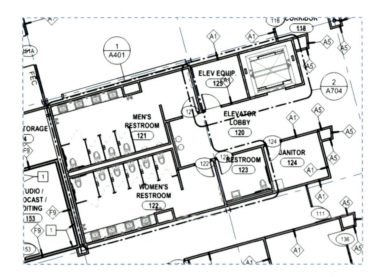

Note that the window on the right will automatically move to match the same location on the page.

* Perform a screen shot and save to your computer.

With the split screen still vertical, you will now open a second file.

7. Click into the left window. From the Menu Bar, select File and then 🗁 Open   Ctrl+O. Navigate to where you saved the sample documents and open the file called, Lab 1 SD Drawings.pdf.

The newly opened file, Lab 1 SD Drawings.pdf should have opened in the left window. The Lab 1 DD Drawings.pdf that was already opened, should now be showing in the right window. It should look like that shown here:

8. From the View Menu, select/check ⬛Synchronize Document

9. In the left window, click on the tab for the **SD Drawings**. In the right window, click on the tab for the **DD Drawings**. Now in the left window zoom and pan to the lower-right-hand corner. Then click the **Next Page** (▷) arrow in the Navigation Bar to advance to the next page. Note that the windows will now move automatically to match the corresponding page and location in each document.

10. Using the Navigation Bar, jump to page 10 of the SD Drawings.

       * *Perform a screen shot and save to your computer.*

11. Now click the unsplit button (⬛) to remove the split screens

Split Window:
Split Window is similar in use case to Split Screen, but is only made possible when using multiple monitors. There are two primary ways to utilize Split Window.

This information is simply for reference. Feel free to try this out if you are working with more than one monitor.

ᗉ When viewing multiple documents, each in a different tab, click-and-drag one of the tabs to the second monitor. This is often referred to as 'undocking' one tab. You will notice that this simply moves the Main Workspace to the second monitor. The rest of the interface and all of the commands remain, but you can undock any panel to move it to the second monitor, if you so desire.

ᗉ The second way to utilize Split Window is through the use of a second instance of Bluebeam. To do this, you would launch another instance of Bluebeam while one is already open. You can proceed to open files within the second instance for Bluebeam by going to **File → Open**.

Profiles

In Bluebeam, Profiles can be used to store the customizable elements of the interface, included which Toolbars are open and where they are located.

Pre-Loaded Profiles

Bluebeam comes with a number of pre-loaded profiles. Follow the instructions below to open and preview four different pre-loaded profiles.

1. We will first select the default profile. From the Menu Bar, select Revu, then Profiles, then Revu as shown to the right. Note that there are minimal Toolbars open.

While we are here, take a moment to open the other pre-loaded profiles (Revu Advanced, Quantity Takeoff, and Field Issues) and note how the interface changes with the settings of each.

- Revu Advanced: Bluebeam would consider this a more robust interface. It is a format that was typical of older versions as well. Note the most commonly used tools located on tool bars across the top of the interface. A key component unique to this profile is the fact that the Panel Access Area is split between the two sides. This may be useful if the user would like to have multiple panels open simultaneously. A larger monitor would be preferable here.

- Quantity Takeoff: As the name would indicate, this profile is geared toward performing takeoff. Notice right away that the Markus List is defaulted to open. A minimal amount of tools, primarily measurement/takeoff tools are located on the right to maximize screen real estate for viewing drawing files. In the Tool Chest panel, the preloaded takeoff toolsets are open.

- Field Issues: This profile is primarily meant for use by folks in the field that are predominantly viewing drawing files and tracking issues/punch items in the field.

2. Make sure to go back to the [Revu] profile before moving on.

Creating Customized Profiles

Now we will create a new Profile and customize some Toolbars. Later in this manual we will modify this profile more.

For our custom profile, we will start with the preloaded Revu profile as a starting point.

1. Confirm you are on the correct profile by going to the Menu Bar, select Revu, then Profiles, then Revu as shown to the right.

2. Now, from the Menu Bar, select Revu, then Profiles, then

3. In the Manage Profiles dialogue box, click Add...

4. Next to Name: type **Real World Bluebeam Profile** and click OK

Add Profile	×
?	Select Name for New Profile
	Name: Real World Bluebeam Profile
	OK Cancel

5. Before moving on, while you are still in the Manage Profiles dialogue box, perform a screen shot at this point.

* Perform a screen shot and save to your computer.

Click OK

6. Go back to Revu, then Profiles to confirm that the newly created
Real World Bluebeam Profile is checked.

Now we'll customize which Toolbars are showing and
where they are located.

7. From the Menu Bar, select the Tools menu and hover over
Toolbars. You will see a list of available Toolbars.

8. Next select Customize...

9. With the Customize Toolbars dialogue box open:
   a. Under Toolbar:, highlight File by clicking on the word
   File. Do not check the box.

   b. Under Items:, confirm that only the following are checked:
   New PDF, Open, Save, Print, Save As. All others
   should be un-checked.

   c. If these are not found under Items:, you may need to add them from the
   list of Commands:, by selecting each and clicking the arrow →

   d. Next, under Toolbar:, highlight Edit by clicking on it. ☐ Edit

   e. Confirm all items are checked for this toolbar.

10. Before moving on, while you are still in the Customize Toolbars dialogue box, perform a screen shot at this point.

* Perform a screen shot and save to your computer.

Click

11. Next, from the Menu Bar, once again select Tools menu and hover over Toolbars. Note that Navigation Bar, Properties Toolbar Shapes, and Text are currently checked. In addition to those, you will also check: **Edit**, **File**, **Measure**, **Order**, and **Rotation**. The top of your interface should be similar to that shown below.

Now we will reposition the additional Toolbars.

12. If you hover your curser over the dots on the left-hand side of each Toolbar (▐), you'll notice a four-way navigation arrow. Click and drag to move the Toolbars around. Keeping all of the Toolbars at the top of the interface, we will ***re-order them*** from left-to-right in the following manner:

File

Edit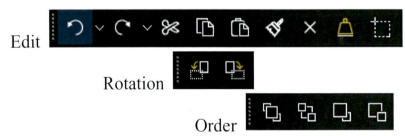

Rotation

Order

The Measure Toolbar should go all the way to the right.

Once complete, your toolbar should look similar to that shown here.

13. Finally, from the Menu Bar, select Revu, then Profiles, confirm the Real World Bluebeam profile is checked, and click Save Profile, to store these configurations.

Perform a screen shot and save to your computer

Exporting/Importing Profiles

You have now created a Custom Profile. The configurations for this profile are saved to a .bpx file, a file type unique to Bluebeam Revu. When a custom profile is created, the .bpx file is saved to the computer hard drive. One can imagine a situation where it may be necessary to share the custom profile with another Bluebeam user. Perhaps a new Estimator or Project Manager has been hired at a firm where a standard profile is used across a department. It may also be necessary if a user switches from one computer with Bluebeam to another, as may very well be the case for a student using their institution's computer labs.

Follow the instructions to export the custom profile from one computer and import it on another.

1. Once again, in the very top-right-hand corner, click Revu, then Profiles, and Manage Profiles

2. In the Manage Profiles dialog box, click on the newly created Real World Bluebeam Profile, and then click Export...

A Save As window will appear. The two most common ways of sharing the .bpx file between computers is to either save it to a flash drive, or to save it to the computer currently be used, and then email it to be downloaded and used on another computer.

3. In the Save As dialog box, navigate to where you would like to save the .bpx file and click Save

You have now saved the .bpx file for use on another computer that has Bluebeam loaded.

If you are using a computer where the custom profile has not been created, or if you wish to share a custom profile with another Bluebeam user, a .bpx file can be imported for use.

Follow the instructions to import the .bpx file.

4. Once again, in the very top-right-hand corner, click Revu, then Profiles, and Manage Profiles

5. In the Manage Profiles dialog box, click on the newly created Real World Bluebeam Profile, and this time click Import...

6. Navigate to where the Real World Bluebeam Profile.bpx file was previously saved, and then click Open

7. In the Manage Profiles dialog box, select the Real World Bluebeam Profile

8. Then click OK

The custom profile has now been added to Bluebeam. You can now close the program.

Lab 2

Document Management

Drawings Management

If it is not already open, follow the directions to open Bluebeam. From the sample documents provided, open the **Lab 2 SD Drawings**.

- Immediately perform a **File → Save As**, or simply select the command button from the toolbar.

- In the File name: field, enter **Real World Bluebeam Revu – (your name) Lab #2 SD Drawings**.

  - It is always a good idea to save your project documents in a dedicated file folder. So browse to a folder of your choice to save.
    <u>*Note:*</u> After you finish saving to the hard drive as described in these instructions, it is ***also*** a good idea to save the file to a flash drive and/or other location as well, ensuring that you have it saved in at least two separate locations.

Thumbnails Panel

The Thumbnails Panel can be opened by selecting the Thumbnails icon from the Panel Access Area on the left of the screen.

1. As we've seen before, the Panel Access Area can be resized. Hover your curser between the Panel Access Area and the Main Workspace until you see the two-sided arrow curser ↔

2. You can resize the thumbnails using the slider at the bottom of the Thumbnails panel

3. See a menu of options by clicking the dropdown arrow next to the word Thumbnails at the top-left of the Thumbnails panel (see right).
OR
Simply right-click on any thumbnail. Note the options available as compared to those in the dropdown previously noted.
   - The most commonly used items in this menu are the typical Cut, Copy, and Paste.
   - We will explore more options shortly.

Page Labels

4. The button next to the Thumbnails dropdown menu is called Labels
Use this dropdown menu to toggle on and off the page labels and page scale. We will add the labels and scales next. Confirm both Page Label and Page Scale are checked as shown here.

Page labels can be created/edited manually by simply double clicking on a page label, or by using the shortcut F2, and entering a name based on your preference.

There is a way to easily do this for all pages automatically though.

5. First, navigate to page 3, G002 CODE INFO, PROJECT GENERAL NOTES, ABBREVIATIONS AND SYMBOLS

6. To create the page labels, select the **Create Page Labels** button at the top of the Thumbnails Panel.

In the **Create Page Labels** dialog box, we can choose to create page labels from Bookmarks (Bookmarks will be covered in a later lab in this manual), or from **Page Region**. The **Page Region** option is most useful to select information from the page's title block to be used in labeling the sheet.

7. For this exercise, select **Page Region**, and then click the **Select** button to setup the page region that will be used to automatically create the page labels.

8. You'll see your curser turn into a cross-hair-style selector to establish the area of the page to be used in creating the page labels. This is referred to as the **AutoMark** function.

9. Select the sheet number from the title block.

   Note: Be sure to select the entire box where text is located as shown with the **blue** rectangle.

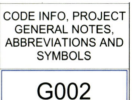

10. You'll then notice that the **AutoMark** box appears with your selection and a preview.

11. Next add another region to the page label by simply adding an underscore (_) after the first selection and then click Add

12. Select the sheet title from the title block .

   Note: Be sure to select the entire box where text is located as shown with the **blue** rectangle.

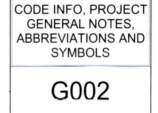

13. Confirm the AutoMark dialog box matches the example to the right and click OK

14. In the **Create Page Labels** dialog box, confirm the **Page Range** is **All Pages**, and click OK

15. Review to confirm all page labels have been created appropriately and manually correct any as needed.

In reviewing the page labels, notice that the first page, the Title Sheet was not named appropriately by the AutoMark tool.

16. Manually correct this sheet to be labeled, **A000_Title Sheet**.

Deleting Pages

Notice that the second page is a blank sheet left in error.

1. To delete this page, first click on the page thumbnail.

2. Then, either select from the Thumbnails dropdown, or simply right click on the page thumbnail and click

3. In the Delete Pages dialog box, confirm the pages selected:

Pages: Selected (2) ⌄ of 14

Reordering Pages

As you continue to review the page labels, notice that the last three pages appear to show A911, followed by A900, and then A912.

1. Click and review each page to confirm that they were labeled correctly.

2. After doing so, we now need to reorder these pages to put them in the proper order.

3. This can be done by simply dragging and dropping the page from one place to another.

4. Note that as you click and hold to drag a page, a **blue** line will appear to show the location of where the page will be dropped.

Inserting Pages

1. Click the Thumbnails dropdown menu, or simply right click anywhere in the Thumbnails panel and notice two commands in particular.

> *Note:* Insert Blank Page may be useful if you are inserting a page as a place holder, or if you plan to create content on the blank page.

> *Note:* Insert Pages can be used to insert a separate sheet or document as a page in the document currently open.

2. Select **Insert Pages**.

3. In the Insert Pages dialog box, click Add (as shown to the right)

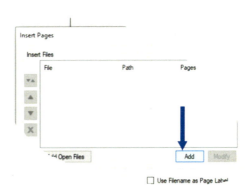

> ***Important Note:*** For some users the Select Files to Insert file selection window will open immediate allowing you to skip step #3 above.

4. Navigate to the sample documents provided and select **Lab 2 Site Concept Plan**.

5. Then click [Open]

6. In the Into. . . section, select **After** from the dropdown, and toggle **First Page**, as shown to the right.

7. Click [OK]

Rotating Pages

Notice that the inserted page is oriented 90 degrees to the rest of the document.

1. Right click on this page and select , or simply use the keyboard shortcut Ctrl+Shift+R.

2. Note that in the Files section of the Rotate Pages dialog box that you can confirm you are in the appropriate file, and have selected the correct page or range of pages. In our case, it should show page (2).

3. Next to Direction, we want to rotate the page **Clockwise 90 degrees**.

4. Then click **OK**

Replacing Pages

Replacing pages can be beneficial for updating a current sheet with a new revision. This is especially useful if a sheet currently has markups or hyperlinks that you would like to retain on the revised sheet.

1. Scroll to sheet A101_FLOOR PLAN -FIRST FLOOR and select it.

2. Right click on the page and select Replace Pages... Ctrl+Shift+Y, or simply use the keyboard shortcut Ctrl+Shift+Y.

3. Navigate to the sample documents provided and select **Lab 2 A101 REV.01**. Then click Open

4. Be sure to confirm that **Replace page content only** is checked.

5. Then click **OK**.

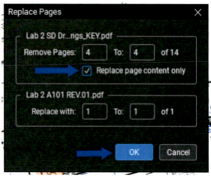

At this point we are finished with the **Real World Bluebeam Revu – (your name) Lab #2 SD Drawings** document. You will submit this document to your instructor along with all other Lab 2 documents.

You should now close this document, but leave Bluebeam open.

To do this, you will find the tab at the top of the Main Workspace. Each individual file that is open will be open in a separate tab and can be closed individually without closing out of Blubeam completely.

6. As this is likely the only file you have open, simply click the white X on the tab for **Real World Bluebeam Revu – (your name) Lab #2 SD Drawings**.

Compare & Overlay

In essence, these two functions accomplish very similar objectives in different ways. Both allow us to easily identify the differences between two documents or revisions.

Compare Documents

Revisions and changes to construction drawings can happen frequently. In Construction Management, it is extremely important to ensure that all necessary parties are informed of any changes to the drawings and that all contractors are working off the most recent revisions.

When revisions are issued by the design team, it is important to understand what has been revised. Relying on clouds placed on the drawings by the design team will not always guarantee that all impacting changes are identified. The **Compare Documents** function in Bluebeam Revu allows the user to easily compare a revised sheet to its superseded sheet, automatically identifying all changes.

1. Navigate to the sample documents provided and select **Lab 2 A112**.

For our purposes, this sheet will be considered a current version that will be compared to a new revision that has been issued.

2. To begin the document comparison, from the Menu Bar, select the **Document** menu, then select **Compare Documents...**

This will open the Compare Documents dialog box. The Document A and Document B sections are for selecting the documents and/or sheets to be compared.

3. The Document A section should be the document currently being viewed. In our case, we will simply view the only page in the document.

4. In the Document B section, navigate to and/or select the second file to be used in the comparison. For our purposes, we are simulating a scenario where the design team has issued a revision to this particular sheet, which we will now compare against the current revision. Click the open the ellipses button (⸱⸱) and navigate to the sample documents provided and select **Lab 2 A112.REV.01**.

5. Then click [Open]

6. Leave the defaults for all other selections and click [OK]

Notice that the split vertical view opens automatically. A new file has now been created with a suffix _Diff which is showing a clouded and shaded area over any differences between the two documents.

* Perform a screen shot and save to your computer.

You may now close the tab for **Lab 2 A112.REV.01**.

Overlay Pages

Now that you have worked with the Compare Documents function, you will use the Overlay Pages function. In essence, these two functions accomplish very similar objectives in different ways. Both allow us to easily identify the differences between two documents. However, with the overlay function, rather than clouded and shaded areas, the differences will be identified by the color of the content itself. When the documents are overlaid, Bluebeam Revu will create new layers for the vector content with new colors. Any place the layers overlap, the content will look ***black*** as it normally would. Where the content does not overlap, the content will be shown in its new color.

Let's begin:

1. If it is not already, open Bluebeam Revu or close any files that may be open.

2. From the Menu Bar, select the Document menu, then select [Overlay Pages...]

3. Navigate once again to the sample documents provided and select **Lab 2 A112** as the first document for the overlay.

4. In the Add Layer dialog box, leave all default settings and click [OK]

5. Next in the Overlay Pages... dialog box, click [Add] to select another document.

6. Navigate to the sample documents provided and select **Lab 2 A112.REV.01**.

7. In the Add Layer dialog box, leave all default settings and click [OK]

8. In the Overlay Pages … dialog box, you should see the two documents to be overlaid. Confirm yours matches as shown here and click OK

9. Note what differences are shown as compared to the Compare Documents function.

 * Perform a screen shot and save to your computer.

10. You may now close the tab being used.

Extracting Pages

Extracting pages from a file will remove pages from the current file and create a new file with them. This can be useful if there was a need to share only a portion of a very large document. The portion can be extracted to create a new document that can then be sent.

Another popular use case for extracting pages is for very large drawing sets. In the case of drawing files that are very large, the files can take a very long time to load and regenerate while scrolling through the file. A popular solution is to break up the large file into its components, such as by drawing discipline. Having established page labels in our file, this can be done very easily.

1. Navigate to the sample documents provided and select **Lab 2 Combined CD Drawings**.

2. Notice that the page labels have already been created. From the Panel Access Area, click the icon to open the Thumbnails Panel (⊞).

3. Right click anywhere in the Thumbnails Panel and select ▭ Extract Pages…   Ctrl+Shift+X , or simply use the keyboard shortcut Ctrl+Shift+X.

4. In the Extract Pages dialog box, in the Page Range section, we could choose to extract an individual page, a range of pages, or all pages. For our purposes, we want to extract All Pages.

5. In the Options section, check the box for **Extract pages as separate files** and then, **Use page label to name files**.

6. Confirm the checkbox for Overwrite existing files is not checked.

It is strongly recommended that you uncheck the box for Open files after extraction in this case.

7. Confirm the dialog matches as shown here and click OK

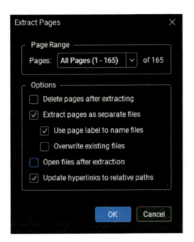

8. Navigate to and create a location you would like to store the individual files where you can remember and access easily.

9. Create a folder called **Lab 2 Extracts**. Then click **Select Folder**.

10. You may then see the warning as shown to the left. This will depend on your operating system and on the sheet titles used as names for the page labels. Simply click OK

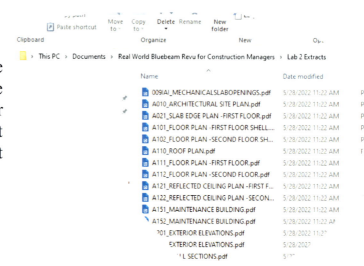

11. Next, from your computer's file management system, navigate to the folder where your extracted sheets are located. It should look similar to that shown to the right.

*** Perform a screen shot and save to your computer*.**

Bookmarks & Hyperlinks

Bookmarks

Bookmarks allows the user to easily jump to an area of the document by navigating through the Bookmarks panel.

1. To begin, if it is not already open, open Bluebeam Revu, navigate to where you had previously saved your **Real World Bluebeam Revu – (your name) Lab #2 SD Drawings**.

2. From the Panel Access Area, select the Bookmarks panel (🔖).

3. You can manually enter individual bookmarks using the Add Bookmarks button and dropdown (🔖⌄). But we will be using the Create Bookmarks function (🔖).

4. After clicking the Create Bookmarks button (🔖), the Create Bookmarks dialog box will appear. Previously we used the AutoMark function to auto-create page labels. These page labels can now be used to easily and automatically create bookmarks.

5. Leave all default selections and simple click [OK]

* Perform a screen shot and save to your computer.

Hyperlinks

Hyperlinks can be extremely useful. There are many ways to utilize hyperlinks in Construction Management, and users continue to come up with creative applications for the hyperlinks functionality in Bluebeam Revu. Here we'll practice just a few examples.

Let's start by using a markup previously created on a sheet to hyperlink to another area of the document.

1. If it is not already open, open Bluebeam Revu, navigate to where you had previously saved your **Real World Bluebeam Revu – (your name) Lab #2 SD Drawings**.

2. On page 4, A101 FLOOR PLAN -FIRST FLOOR, notice a magenta colored square in the center of the page over an elevation marker labeled A201 as seen to the right.

3. Right click on the object and select `Edit Action...    Ctrl+Shift+E`

4. In the **Action** dialog box we can select from a number of ways to use this markup object as a hyperlink. For this first example, we will hyperlink to another page within the document in which we are currently working. Select the options shown here and click `OK`

5. Notice that when you click on the small lightning bold near the magenta box, it will now take you directly to that page.

In this next example we can jump to an even more specific area within the document currently being viewed.

6. In the same area as the markup previously used on page 4, A101 FLOOR PLAN -FIRST FLOOR, this time right click on the **red** circle around the number (**7**), and once again select `Edit Action...   Ctrl+Shift+E`

7. This time, select the **Snapshot View** option, and click `Get Rectangle` . At this point you will navigate within the document to page **9** and use the rectangle selection, click and drag, to select an area as shown here, as indicated by the **blue** rectangle from elevation **7** on sheet **A201**.

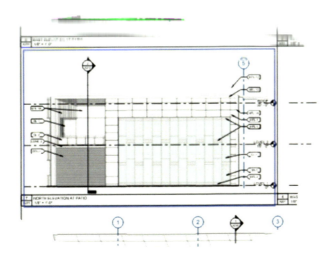

Now, we will hyperlink to a web address outside of our current document.

8. Navigate to page 12, A900 FINISH LEGEND AND ROOM FINISH SCHEDULE. Right click on the orange cloud object and once again select `Edit Action...   Ctrl+Shift+E`. This time, select the **Hyperlink** option and enter the following text into the text field: **https://roppe.com/pinnacle-rubber-base/**.

9. Click `OK`

Note: Hyperlinks can be created from the Links panel (📍), as well as my going to the Menu bar to select the Tools menu, and from there selecting `🔗 Hyperlink   Shift+H`

10. At this point be sure to save your work to submit to your instructor before closing the document.

File Management

Studio

Studio in Bluebeam Revu serves as a basic file management system, allowing users to house and share files in the cloud. Studio has a check-in/check-out function that helps to ensure proper versioning for document integrity.

1. If it is not already, open Bluebeam Revu and select the Studio panel (⬡) from the Panel Access Area.

2. In the Studio panel, you will first need to select a server from the dropdown menu (shown right) and then click **Sign In**. You will then be prompted to either sign in or create an account. Create an account you will use for Bluebeam Studio.

Once signed in, from the Studio panel, you can access and create Studio Projects. Projects in Bluebeam Studio are a way to organize files together as a related to a project. Collaborators can be invited and managed as needed and as related to that project.

3. Click the Add button (➕) and then 📄 New Project to create a new project.

4. Name your project **Real World Bluebeam Project Documents,** as shown here and click OK

5. You now have the ability to add all documents using options to either create a new folder or upload files/folders. Start by clicking **New Folder**, and create a folder called **Project Drawings**.

6. Double click on the Project Drawings folder and then select

7. In the Upload Files dialog box, from the dropdown, select **Files**.

8. Navigate to the sample documents provided and select **Lab 2 SD Drawings** and **Lab 2 Combined CD Drawings**.

* Perform a screen shot and save to your computer.

Lab 3

Tools

Markups

Bluebeam Revu offers a variety of markup tools that can be used to annotate and communicate visually in documents. Bluebeam offers a set of markup tools that can be used in a variety of ways, depending on the application. We will review each in turn.

If it is not already open, follow the directions to open Bluebeam. From the sample documents provided, open the **Lab 3 CD Drawings**.

- Immediately perform a **File→Save As**, or simply select the command button from the toolbar.

- In the File name: field, enter **Real World Bluebeam Revu – (your name) Lab #3 CD Drawings**.

    - It is always a good idea to save your project documents in a dedicated file folder. So browse to a folder of your choice to save.
    Note: After you finish saving to the hard drive as described in these instructions, it is *also* a good idea to save the file to a flash drive and/or other location as well, ensuring that you have it saved in at least two separate locations.

1. First navigate to the second page, sheet A111, FLOOR PLAN FIRST FLOOR.

2. From the Menu Bar, click Tools and hover over Markup to see the options of markup tools as shown here. Note that many of the tools in this list are also available on the toolbars previously added in Lab 1.

Text Box

3. The first two tools in the list are Text Box and Typewriter, which are very similar in that they will insert text as an annotation. Click [A] Text Box   T

4. You'll find your cursor changes [A] Click and drag to create the box in which you will add text. Using your mouse wheel to navigate, place the text box in the TRAINING 3 room 163, and then type the following: **Movable chairs to be used for flexible room arrangement.**

In the top-left-hand of the interface, just below the File toolbar, will appear the Properties toolbar, containing some of the most common customizable properties for the object.

You can also find the full options for properties in the Properties Panel by selecting the properties icon [⚙] from the Panel Access Area.

5. Change the Line Color to **Dark Green** and Line Width **3.00**. Then, set the line Style to [Dashed 3   - - - - - -]

6. Set the Fill Color to Green with a Fill Opacity of **30**.

7. Click and drag to highlight the text in the box and change the Text Color to **Black**.

8. Click the Autosize Textbox button ([⊡])in the Properties toolbar.

9. Your Text Box should look similar to that shown here.

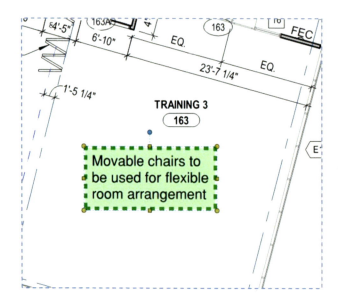

Typewriter

The next markup tool in the list is the Typewriter tool ( ). While this is similar to the Text Box tool, both will create a box with text, it is created in slightly different way. Whereas with a Text Box, you had created a box first, constraining your space to type text, with the Typewriter tool, you will simply click anywhere you would like to begin typing.

10. Select the Typewriter tool and then using your mouse wheel to navigate, click in the SHIPPING/RECEIVING room 152.

11. Begin typing the following: **Storage Shelving**.

12. Press the Enter key, and continue typing: **ULINE item H-8613**.

13. Press the Enter key, and continue typing: **2 rows of 2 units each**.

14. From either the Properties Toolbar in the top-left, or from the Properties Panel, change the Line Color to **Dark Blue** and Line Width to **3.00**. Then, set the line Style to Dashed 3 -------

15. Set the Fill Color to Cyan with a Fill Opacity of **30**.

16. Click and drag to highlight the text in the box and change the Font Size to **10**.

17. Resize and/or reposition the text box as needed so that all words are visible. Your text box should look similar to that shown here.

Note

A Note can be useful if the annotation desired is longer than you care to include in a box (i.e. pages from the specs, or building code language). With a note, a small note icon ( ) will be shown on the document, with a text field appearing only when the icon is clicked.

18. Click on the Note tool ( ) and click to place the note in the STUDIO/PODCAST EDITING room 153.

19. In the Note text field, type: **Need to insert specifications for technology equipment requirements here**.

Callout

A Callout is comprised of an arrow leader and a text box. An extremely common markup item in the construction industry, it is useful as the arrow can be used to clearly identify the subject of the annotation with the text providing clear explanation.

20. Click on the Callout tool ( ).

21. Locate the column at the intersection of column lines 9 and H. Click on the right edge of the column to locate the head of the arrow. Then move your curser down and to the right, toward the center of OPEN OFFICE room 180, and click again to locate the text box. In the text box type: **Need requirements for column coating and fire rating requirements.**

22. Hit the Esc key and then click on the Callout object. Note that there are a number of grips you can use to resize and reposition both the text box and the arrow.

23. Right click on the Callout object, and select [🖊 Add Leader]. Place the new arrow leader on the column at column lines 10 and H.

24. Reposition the arrow leaders such that both connect to the right side of the text box.

25. Using the Properties panel, change the Line Color to **Red Orange**, the Fill Color to **Yellow**, Fill Opacity to 30, and the Line Width to 2.00.

26. Your callout should look similar to that shown here.

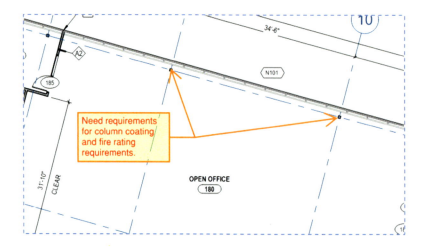

Pen, Highlight, Eraser

The next family of markup tools are grouped together including Pen, Highlight, and Eraser. The Pen tool can be used freely using either a mouse with a left-click-and-hold, or on a touch-screen. Hold the Shift key to draw a straight line either vertical or horizontal. The Line Color, the Line Opacity, and the Line Width can all be customized. The Highlight tool works in all the same ways as the Pen. In addition, where text is recognized in the PDF, the cursor will automatically change for you to select the text to be highlighted.

2.7. In the KEYNOTES – FLOOR PLANS, use the Highlight tool to highlight #3, 3 | 1-HR RATED FRAME AND GLAZING . Press Esc. Then select the highlighting and change the color to Pea Green.

Shapes

The next family of markup tools are in a group we'll call Shapes. In this family of tools, we find many of the conventional shapes traditionally used in PDF markups: Line, Arrow, Arc, Polyline, Rectangle, Ellipse, and Polygon. We'll address Dimension separately. For all of these shapes, we can customize the Line Color, the Fill Color, the Opacity, the Line Width, and the Line Style.

- Line & Arrow: Note that the Line tool includes Line Start and Line End customization, making it nearly identical to the Arrow tool.

  - For the Arrow tool, click and drag in the direction you would like the arrow to point.
  - Hold the Shift key while drawing these shapes to create straight lines in cardinal orientations.

- Polyline is used to create multiple connected line segments.

  - Click to begin the polyline, and then click again at each vertex of the shape. Double click to complete the shape.

- Rectangle & Ellipse: Click and drag create the shape.

  - Hold the Shift key to create a perfect square or circle respectively.

The Dimension tool is a little different. It takes on similar characteristics as a line or arrow, but also includes text for its label. While this may look like a measurement, it is simply a markup, and has no scale properties associated with it.

28. Click on the Dimension tool.

29. To start the dimension line, click on the left wall of TRAINING 1 room 161, then click on the right wall of TRAINING 3 room 163. Then type **89' – 0 ¼"**

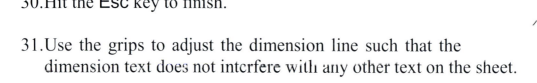

30. Hit the Esc key to finish.

31. Use the grips to adjust the dimension line such that the dimension text does not interfere with any other text on the sheet.

Cloud & Cloud+

The Cloud and Cloud+ tools are extremely popular in the construction industry, particularly on drawings. The Cloud tool can be used to draw a cloud around PDF content.

> *Note:* The Line Style of a Text Box can be made to match that of a Cloud in order to create the look of a Cloud, but containing text.

The Cloud+ tool combines a cloud with a leader and a text field. For both Cloud and Cloud+, all properties are customizable the same as previously covered.

32. Click on the Cloud+ tool (⌨) and place a cloud around the stairs in the center of the Main Lobby room 111. Then click to place the leader and text field up and to the left in the blank space.

33. In the text field, type: **Stairs not included in the Structural Steel work package.**

34. Hit the Esc key to finish.

35. Use the blue circle rotation grip (*identified at right*) to rotate the cloud counter-clockwise to match the orientation of the stairs. Then use the blue circle rotation grip at the text field to rotate it clockwise so it is back to being horizontal in orientation.

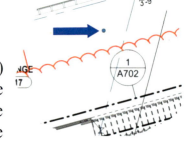

36. Your **Cloud+** should look similar to that shown here.

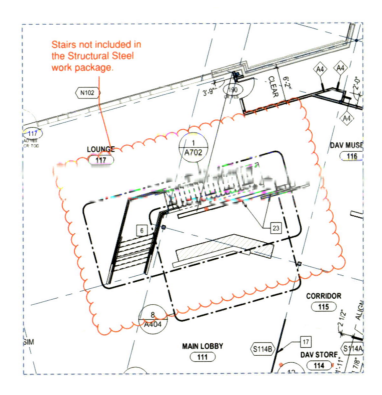

Measuring & Takeoff

Arguably one of the greatest selling features of Bluebeam Revu is its ability to take scaled measurements and use this to do takeoff or quantity survey of the drawings. Remember that a measurement is nothing more than a markup with additional properties associated with it. All of the same customizable properties covered for markups can also apply to measurements. We'll now practice each of these tools in turn, and then see what we can do with the results.

1. If it is not already open, follow the directions to open Bluebeam. Open your **Lab 3 CD Drawings** as you left them following the **Markups** exercises.

2. All of the markups we practiced were done on the second page. You'll now navigate to the third page, A112 FLOOR PLAN – SECOND FLOOR.

Recall that we had previously taken steps to customize the interface and created a new Profile for use throughout this lab manual. We will now make some small modifications to one of the Toolbars placed on the interface.

3. Confirm you are still on the correct profile by clicking in the very top-left-hand corner on **Revu**, then hover over **Profiles** and confirm you are on the Real World Bluebeam Profile, indicated by the checkmark.

4. Next we will add tools to the Measurement Toolbar. From the Menu Bar, click on Tools, then click Toolbars and go down to click on Customize...

5. In the top-right of the Customize Toolbars dialog box, there is a field called Toolbar: . From this area, click on the word **Measure**.

6. Note that all items are currently checked in the Items: area. We will add others now. In the top-left, under Categories: , select **Measure** from the drop down menu. Then, under Commands: select the following by clicking on the command, and then the →:

- 3-Point Radius

- Center Radius

- Diameter

- Dynamic Fill

- Radius

7. Finally, click OK

Measurement Tools

Sometimes we don't want to perform a quantity takeoff for specific pieces of work, but we need to take some simple measurements for various reasons. Let's take a look at some of those basic measurement tools.

Before doing any measurement or takeoff, we must establish the scale for the page. There are two ways to establish the scale. We will try both.

Calibrate

8. From the **Panel Access** Area, open the **Measurements Panel** ( ). Then next to **Scale**, click Calibrate

9. A window of instruction will appear, directing you to click on the two points of a known dimension – that is, a dimension that is labeled on the drawings with which you can establish a scale and against which you can check your scale. It is highly recommended to use the longest dimension available for better accuracy. Let's use the dimension line between column lines C and D. After clicking on the end points of the dimension line, the Calibrate dialog box will appear. A field is automatically engaged in which you will enter to the length verbatim as **44'-0"**. Leave all other fields as default.

Enter Scale as Labeled

With modern drawing sets plotted from a CAD software, the scale as labeled is most likely reliable and can simply be entered for use.

10. With the **Measurements Panel** ( ) open, below **Scale**, click the ⦿ Preset radio button. Then from the dropdown selection, click the scale for your sheet.

11. Notice in the very bottom-left-hand corner of our sheet, it is labeled:

| 1 | FLOOR PLAN - SECOND FLOOR |
| A112 | 3/32" = 1'-0" |

From the dropdown selection, click 3/32" = 1'-0"

Note: In earlier versions of the Bluebeam software, the preset options for commonly used scales were not available, but simply provided a field to input the scale.

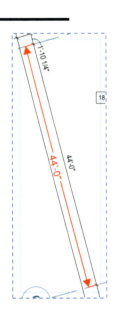

Length

A common use of the Length measurement is to double check the scale as entered against a known dimension on the sheet.

12. From the Measure Toolbar, select the **Length** (▣) tool.

13. Find the dimension line previously used to calibrate the scale, between column lines **C** and **D**. Click once at one end of the dimension line, and then again at the other end. The length measurement should appear in red as shown here to the right.

Angles

14. Near the center of the page, there are two angles labeled ⌐112.00° Two walls meet to create this angle. To check this angle, first from the **Measure Toolbar**, select the **Angle** (◣) tool.

15. You will need to click in three places in a sequence to measure this angle. Any angle is created by two lines intersecting. Therefore, to measure the angle, we can identify a point anywhere along the first line, then at the point of intersection, and finally a point anywhere along the second line. For our example, click first along one of the walls intersecting, then at the point where they intersect, and finally a point along the second wall. See below.

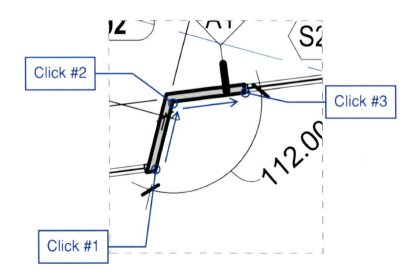

Diameter

For the next two items we will navigate to page 13, sheet C1.3 MATERIALS PLAN – ENTRY & EAST PLAZAS.

16. Start by setting the scale for the sheet. Similar the steps previously covered, we will enter the scale as shown: 1" = 5'. This time we will click the radio button for ⬤ Custom , because the scale we need is not one of the presets.

17. Type in the scale to match as shown below:

18. On the right half of the sheet, there is a circle with a note numbered 5 in the middle. Click on the Diameter (⊙) tool.

19. Next click anywhere along the edge of the circle. As you move your curser to another point along the edge of the circle, you will snap to the circle. As you follow the edge of the circle, letting your curser continue to snap along the line, you can move your curser until your measurement matches the circle being measured.

20. Click again to complete the measurement.

Radius

We will use the same circle to practice a radius measurement. You may have noticed that there are two different types of radius measurements. Thc simpler of the two is the Center Radius tool. If the center point of the arc/circle is identified, you can simply click on the center point and then the edge to find the radius. Perhaps more common though, is finding a radius when the center point is not shown.

21. Click on the 3-Point Radius (⊙).

> *Hint:* The icons for 3-Point Radius, Center Radius, Radius, and sometimes others, can look very similar, especially on a small screen. You may need to hover your curser over the icon to confirm you are selecting the correct one.

22. Click anywhere along the edge of the circle, then click another point anywhere along the edge of the circle.

23. You will notice that a blue circle forms, and as you move your curser, it will move. Now you will want to move your curser along the edge of the circle until the blue circle matches that which you are trying to measure. Click again to complete the measurement.

Count/Each Takeoff

The count takeoff tool, often referred to by its unit, being called an "Each" count or takeoff, is used to simply count a number for any takeoff item.

To practice this tool, we will navigate once again to page 3, A112 FLOOR PLAN – SECOND FLOOR. On the left-hand side of this sheet, there is a row of rooms consisting of 225, 226, 235, 236, S12, 237, 238, 239. We will count the number of doors to these rooms.

24. Click on the Count tool (▦).

25. From the Panel Access Area, click on the Properties Panel (⚙).

26. Near the top of this panel is a section called General. In that section is a field called Subject:. You can think of this field as a name for the takeoff you are performing. By default, the subject will be the name of the takeoff tool being used. For the most useful results, we need to name what it is we are going to takeoff, so that it can be referenced later. In this case, we are taking off a group of doors. For our purposes, we will simply name this takeoff as Doors. In the Subject: field type the word **Doors**.

    <u>*Note*</u>: If we were to leave this takeoff by clicking Esc., or work on other takeoff items, and then wish to restart this Doors takeoff, we can simply restart the Count tool, and again type the appropriate name in the Subject field.

27. In the Appearance section, we can customize the appearance of the annotation associated with this takeoff item. For our purposes, we will leave the default settings.

28. Begin clicking on each door opening. It is recommended for any takeoff item to start in one place on the sheet and proceed systematically in order to ensure that no items are missed. We will begin at the bottom of the sheet at door (239) and proceed upward.

29. Click on (239), (238), (237), and (S12), and then hit Esc.

We are not finished with the takeoff we intended to complete, but we have now closed the Count takeoff that we started. We have simulated a situation where the takeoff was mistakenly closed prematurely, or perhaps something pulled you away from your work in the middle of a takeoff. Often you'll simply notice an item you had missed. Let's now see how we can continue with a takeoff that has been closed.

30. Right click on any of the Count takeoff items already done, and click `Resume Count`

31. Notice that the subject once again says Doors before proceeding.

32. Continue the takeoff by clicking at (236), (235), (226), and (225). Then click Esc. to close the takeoff.

Linear Takeoff

Length

Similar to a simple length measurement, the Length tool can be used multiple times, in a number of individual locations, and combined to create a cumulative takeoff of various individual lengths.

Polylength

With the Polylength takeoff tool, we will takeoff a series of individual, but connected, length measurements at one time. This can be very useful for piping.

33. Navigate to page 17, sheet **P-02: Level 02**. In the center of this page, there is a pink colored section of pip labeled **6" OFD** (see below).

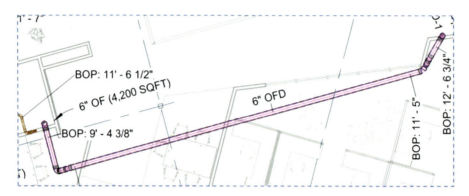

34. Again set the scale of the sheet in the **Measurements Panel** (⬜) under **⊙ Preset**

35. Begin the takeoff of this section of pipe. Select the **Polylength** tool (⬛).

36. From the **Panel Access** Area, click on the **Properties Panel** (⚙).

37. Near the top of this panel is a section called **General**. In that section is a field called **Subject:**. We will name this takeoff for the pipe being taken off. In the **Subject** field, type **6" OFD**.

38. Click at one end to begin the takeoff. For pipe, it is best to follow the centerline of the pipe for takeoff. Zoom in using your mouse wheel to be as precise as possible. Click at each intersection of segments and then double click to close the takeoff. Then hit **Esc**.

___

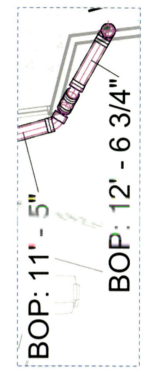

Note: There is a portion of the pipe near the right edge of our takeoff that may be somewhat confusing because there is a section of pipe that is vertical in orientation, which is difficult to discern on a 2D drawing sheet (see right). Notice the designer has indicated the bottom of pipe elevations (BOP) throughout the sheet. In this area, the pipe changes from a BOP elevation of 11' – 5" on the right section to 12' – 6 ¾" on the left section of this segment. A change of 1' – 1 ¼" in elevation.

At the left-hand side of this length is another segment that is vertical in orientation. It goes from 9' – 4 ⅝" to 10' – 11 ¼", a change in elevation of 1' – 6 ⅝". The total rise/drop of pipe therefore, is 2' – 8 ⅝".

39. Click on the Measurements Panel (🗏) to add a Rise/Drop length of **1' – 1 ¾"**.

Note: If you click on this Polylength takeoff, you'll notice that the overall length is labeled, along with the individual lengths of each segment of the overall length. If necessary, these segments can be separated, one from another adjacent segment, or all separated completely by simply right clicking on the item, and selecting either Split or Split All respectively. New properties can be associated with each segment individually if needed.

Perimeter

The Perimeter tool is similar to Polylength, but is intended to take off the perimeter of a closed shape, such as a room, and has some additional properties that can be useful.

40. Navigate once again to page 3, A112 FLOOR PLAN – SECOND FLOOR. Near the top-right-hand corner of the sheet is room 292 BOARD ROOM.

41. From the Measurement Toolbar, select the Perimeter tool (▱).

42. Once again, from the Panel Access area, click on the Properties Panel (⚙).

43. We will name this takeoff by typing **Wall Area** into the `Subject:` field.

44. To begin the takeoff, click in one corner of the room and continue clicking in each corner around the room. Double click to close the takeoff. Then click Esc.

   ***Note*:** Yellow grips will appear at each corner that can be used to adjust the takeoff as needed.

45. Click on the Measurements Panel (📏) to add some additional information that can be useful.

46. In the `Depth:` field we can add a wall height. Type **10'-0"**.

Now from a single Perimeter takeoff, which we named Wall Area, we have data for the perimeter of the room as well as wall area of the room.

Area Takeoff

Area

The Area measurement tool can be used to takeoff the area of a horizontal surface in plan view or a vertical surface in elevation view.

47. Let's now navigate once again to page 11, A911 FINISH FLOOR PLAN – FIRST FLOOR. Near the center of the sheet is room 150 Storage. We would like to takeoff the flooring for this room. According to the legend, this hatch pattern indicates Sealed Concrete.

48. From the Measurements toolbar, click the Area tool (⬚).

49. Again set the scale of the sheet in the Measurements Panel (📏) under `⊙ Preset`

50. From the Panel Access Area, click on the Properties Panel (⚙).

51. We will name this takeoff by typing **Sealed Concrete** into the Subject: field.

52. Similar to the Perimeter tool, you will click in one corner of the room and continue around the room, clicking in each corner. Double click to complete the takeoff.

53. So that we can easily identify this takeoff visually, make the following changes in the properties.

   - Change the Line Color and Fill Color to Magenta
   - Change the Fill Opacity to 30%

Polygon Cutout

The Polygon Cutout tool will create a space within a polygon created by an Area takeoff, thereby removing some quantity from the shape. Let's suppose hypothetically that within the shape of our Sealed Concrete takeoff in room 150 Storage, there is an area of VCT for a maintenance workspace. We can use the Polygon Cutout tool to quickly takeoff this area within another area.

54. From the Measurements toolbar, click the Polygon Cutout tool ().

55. Near the center of the room, click and drag a rectangle shape of any size. Notice that the cutout shape has now been created within the room. You can now get an area for the cutout shape and the area for the overall shape has been adjusted.

Volume Takeoff

A common application for the Volume Takeoff tool is to get concrete volumes for a slab.

56. Navigate to page 20 of this document, sheet S101 LEVEL 1 SLAB AND FOUNDATION PLAN.

57. To begin the takeoff, from the Measurements toolbar, click the Volume tool ( ).

58. Again set the scale of the sheet in the Measurements Panel (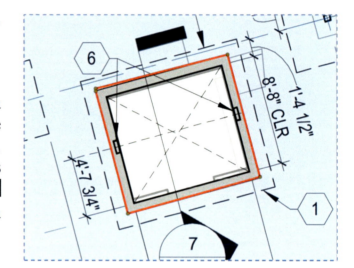) under
   ○ Preset

59. From the Panel Access Area, click on the Properties Panel (⚙).

60. We will name this takeoff by typing **5" SOG** into the Subject: field.

61. As with the Perimeter and Area takeoffs, we will click in one corner and continue around the shape, clicking in each corner, following along the inside face of the foundation wall. For this takeoff, we will takeoff the entire footprint of the building.

62. Click the Polygon Cutout tool (▣), and click to trace the four corners of the elevator pit.

63. Now, we need to indicate a thickness, or depth for the concrete slab in order to get the volume. Click on the Measurements Panel (▦) and find the Depth: field. From the dropdown, select a unit of inches (in).

64. In the Depth: field we can add the slab thickness in inches. Type **5**.

Notice that we can adjust the units for various elements found in the Measurements Panel. For Volume:, adjust the units to cubic yards (cu yd) by selecting from the dropdown.

65. So that we can easily identify this takeoff visually, make the following changes in the properties.
   - Change the Line Color and Fill Color to **Dark Green**.
   - Change the Fill Opacity to **30%**.

Dynamic Fill

This may seem like the greatest time-saving feature in the Measurements Toolbar, Dynamic Fill. Dynamic Fill can be used to quickly and easily takeoff large and oddly-shaped areas.

66. Let's navigate once again to page 11, A911 FINISH FLOOR PLAN – FIRST FLOOR, where we began to takeoff flooring. If we wanted to takeoff the area hatched as CONC-1, this would take many tedious clicks and steps. With Dynamic Fill, this can be done in a few short steps.

67. From the Measurements Toolbar, select Dynamic Fill (⊞). Notice that a floating toolbar appears.

68. In this toolbar there are a couple selections that are especially important. The first is establishing boundaries. In the Define section the Boundaries button (⊞) can be used to create a boundary in any location where you may need to close the shape to be filled. Because there is a separation shown with a solid line between all flooring types, our overall shape is closed in this case. Therefore no boundaries are needed.

69. The next important setting is Fill Speed, found under Dynamic Fill Settings (shown below).

As you drag the filler throughout the space, it will fill automatically. Until you are comfortable with the tool, it is best to keep the Fill Speed relatively low.

70. Click on the Fill button (⬗), and then begin to click and drag on the drawing in the areas hatched as CONC-1. If the fill stops at a closed shape in the drawings, continue to click until all the desired areas have been filled.

71. Once we have filled all of the area desired, we can now use this selection for various purposes. Notice the other selections in this tool bar next to **Create:** For out purposes, in order to takeoff the area of flooring for this selection, we will click Area (⬚), and then hit **Apply**

72. Then, from the Panel Access Area, click on the Properties Panel (⚙), and name this takeoff by typing **CONC-1** into the **Subject:** field.

73. On the floating toolbar, you can now hit **Close**

74. So that we can easily identify this takeoff visually, make the following changes in the properties.

   ▪ Change the Line Color and Fill Color to Pastel Cyan
   ▪ Change the Fill Opacity to **30%**

Markups List

Up to this point in this manual, there is one area that was referenced in the areas of the interface, but we have not yet used. That is the Markups List.

↻ In the very bottom-left-hand corner, ☰ click to open the Markups List.

↻ Hover your curser between the top of the Markups List and the bottom of the Navigation Bar to resize the Markups List.

↻ You can also undock the Markups List to move it around freely, or even move it to a second monitor to maximize screen real-estate.

The Markups List contains a detailed record of the markups of a document, including measurement and takeoff markups. This information is in various columns that can be customized and configured for various purposes.

Working with Columns

One of the many properties that can be assigned to a profile are columns in the Markups List. Recall that we previously created a profile for this manual.

1   Confirm you are still on the correct profile by clicking in the very top-left-hand corner on **Revu**, then hover over **Profiles** and confirm you are on the Real World Bluebeam Profile, indicated by the checkmark.

2. In the Markups List area, columns can be configured by clicking the dropdown next to Markups List and selecting **Columns** as shown here. The checkmarks indicate which columns are currently turned on. We will now adjust which columns we'd like to see.

3. Click to uncheck each of the following: Author, Date, Status, Color, and Space.

4. Then click to check (turn on) the following: Checkmark, Wall Area, Volume, and Rise/Drop.

   Note: The Checkmark column creates a simple checkbox for each line. This checkbox can be used however the user sees fit. One example might be for a reviewer to check off items as she/he review takeoff performed by a colleague.

5. Confirm you have the following columns checked: Subject, Page Label, Comments, Checkmark, Length, Area, Wall Area, Volume, Count, Measurement, Depth, Layer, Rise/Drop.

6. To reorder to columns, you will simply click and drag the column header to move it to the desired location. Click and drag the columns to place them in the following order: Subject, Checkmark, Measurement, Length, Area, Wall Area, Volume, Count, Depth, Rise/Drop, Layer, Page Label, Comments.

Note: Columns can be used to search, sort, and filter data within the Markups List. Click on a column header to sort by that column's data. Use the 🔽 **Filter List** functionality to filter by various data points depending on the data being viewed in the Markups List.

With the Markups List still open, perform a screen shot

* Perform a screen shot and save to your computer.

Export

The Markups List holds a lot of valuable information that can easily be exported for use in other programs, such as Excel. One common application of this functionality is to complete takeoff in Bluebeam and then export the quantity data to be used in an enterprise estimating software, or simply export to Excel for further manipulation in a legacy spreadsheet.

7. With the Markups List still open, click the Summary button (📤) at the top of the Markups List area, and then select **CSV Summary**.

8. In the Markup Summary dialog box, choose the destination where you would like to save the export file in **Export to:**

9. In **File name:**, name the file **Real World Bluebeam Revu – (your name) Markups Export**.

10.Click the radio buttons to turn on Markups & Totals.

11.Leave all other defaults, and then click **OK**

Link to Excel

Bluebeam includes a feature called Quantity Links, which allows the user to link cells in an Excel spreadsheet directly to the data input source from the document in Bluebeam. This means that when a measurement or takeoff is performed in the document in Bluebeam, that quantity can be automatically updated in the respective spreadsheet.

12. Navigate to the sample document provided to open the spreadsheet named Lab 3 Quantity Link Spreadsheet. Notice that the items listed in the spreadsheet match those used in practice takeoff in previous steps

13. Immediately perform a **File→Save As** or simply select the command button from the toolbar.

14. In the File name: field, enter **Real World Bluebeam - (you name) Lab 3 Quantity Link Spreadsheet**.

15. To create the first link, right click in the Material QTY cell for 5" SOG and hover over Quantity Link.

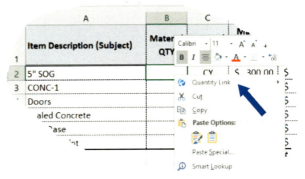

16. As this is the first time creating a link, you will be prompted to create a new link. Click ● Create...

17. Click on the dropdown arrow next to Add, select **Add Files**, and then navigate to select your Lab 3 CD Drawings.

18. Click Open

19. Then select your file from the list and then click OK

20. You should now see the Create Link dialog box where you will link the specific data from your takeoff to this specific cell in your spreadsheet. Next to ^{Total:}, select Volume, as this is the column where the total will be found for this particular item. Where you see [⌐], hover over Subject and then click `5" SOG`, and then click `OK`

21. Continue to link each of the Material QTY cells for each of the other material takeoff items. Keep in mind, the totals for each item will be found in a different column from the Markups List in Bluebeam. This means that next to ^{Total:}

- For CONC-1, select **Area**
  - For Subject, you will select **CONC-1**
- For Doors, select **Count**
  - For Subject, you will select **Doors**
- For Sealed Concrete, select **Area**
  - For Subject, you will select **Sealed Concrete**
- For Wall Base, select **Length**
  - For Subject, you will select **Wall Area**
- For Interior Paint, select **Wall Area**
  - For Subject, you will select **Wall Area**

The data from the Markups List has now been linked to this spreadsheet.

Legends

A legend is a quick way to show a snap shot of information from a given area. Legends can be used to create a table of dynamic data from all markups and takeoff on a sheet or portion of a sheet.

1. Let's navigate once again to page 3, A112 FLOOR PLAN – SECOND FLOOR.

2. On the keyboard, type Ctrl+A to select all of the markups on this page.

3. Right click on any of the markups and then hover over [Legend], and then select Create New Legend

4. Click to place the legend in the open space at the top-center of the sheet.

5. As before, the Properties bar appears at the top of the interface. From the Properties bar, make the following changes:

   - Change the Title to **Takeoff Items**
   - Change the Symbols to **200%**
   - Make the Line color **Black**
   - In the Table Style dropdown select **Gridlines**
   - Change the Line Weight to **1.00 pt**
   - Change the Fill to **Blue**
   - Change the Fill Opacity to **30%**

Layers

Layers can be created to organize content in Bluebeam. Various content elements can be assigned to layers allowing the user to turn content on and off from view. This can be useful in takeoff, as takeoff items can be assigned to layers in order to be turned off from view when not in use. Takeoff items can also be categorized in layers (i.e. framing, plumbing, electrical, etc.).

1. Open the Layers panel from the Panel Access Area by clicking the Layers icon (⊗).

Note that there are many layers associated with the drawings that are viewable from the original authoring process. For now we will disregard these layers, and add different layers for our purposes.

2. Now we will add layers that can be used to organize our takeoff. In the top-left-hand corner of the Layers Panel, click the Add New Layers button (◇⁺ˇ) and select Add Before...

3. In the Layer: field, type **Concrete**.

4. The click OK

5. Repeat this process to create a layer for: **Flooring, Doors, Wall Base, Interior Paint**.

6. If it is not already open, also open the Markups List by clicking the Markups icon (☰).

7. Right click on the first takeoff item 🔲 5' SOG

8. Click **Layer** and then select 👁 Concrete

9. Repeat this process to add each of the following takeoff items to their respective layers:

- CONC-1 → Flooring
- Doors → Doors
- Sealed Concrete → Flooring

10. With the Markups List open, you will now perform a screenshot.

 * Perform a screen shot and save to your computer.

Tool Chest & Tool Sets

In the Tool Chest, various Tool Sets can be used to organize customized tools that can be easily accessed for repeated use, saving the effort of reconfiguring properties for a tool. This can be useful for takeoff tools. Each time we performed a takeoff for a particular item, we configured its appearance and some other properties. Rather than reconfiguring these settings each time we use the tool, we can save the configurations as a customized tool in the Tool Chest.

1. Confirm once again, that you are still on the correct profile by clicking in the very top-left-hand corner on **Revu**, then hover over **Profiles** and confirm you are on the Real World Bluebeam Profile, indicated by the checkmark.

2. From the Panel Access Area, select the Tool Chest icon (🗄).

3. From the Tool Chest ⌄ dropdown, select Manage Tool Sets

4. In the Manage Tool Sets dialog box, click Add...

5. In the ^{Title:} field, type **Takeoff Tools**.

6. Confirm ☐ Show In All Profiles is not checked and click ⟦ OK ⟧ , then click ⟦ Save ⟧ and the ⟦ OK ⟧ again.

7. In the Tool Chest, our newly created Takeoff Tools tool set can now be seen at the bottom.

8. From the Markups List, click on ⟦ 5" SOG ⟧. Then in the Main Workspace, right click on the takeoff item.

9. Click on ⟦ Add to Tool Chest   › ⟧ and then ⟦ Takeoff Tools ⟧

10. Confirm that this item has been added to the Tool Set as seen here.

11. In the Tool Set, double click on the tool to change it from Drawing Mode into Properties Mode.

When the tool is in Drawing Mode, it will be duplicated verbatim with each usc. This is useful if we've created a custom symbol. For our purposes here, we prefer Properties Mode. We would like to duplicate all properties of the custom takeoff tool.

Notice that the icon has now changed

12. Repeat these steps to add the following takeoff items to the Tool Set: CONC-1, Sealed Concrete, Wall Area.

13. Make sure to save your work before moving on.

At this point you are finished with Lab 3. You will submit this document to your instructor along with all other Lab 3 documents.

Lab 4

Field Use

Site Logistics

In Lab 3 Tools, you learned about some of the many Markups tools available in Bluebeam. An extremely common application of these tools is in creating a visual Site Logistics Plan.

If it is not already open, follow the directions to open Bluebeam. From the sample documents provided, open the **Lab 3 CD Drawings**.

- Immediately perform a **File→Save As**, or simply select the command button from the toolbar.

- In the File name: field, enter **Real World Bluebeam Revu – (your name) Lab #4 CD Drawings**.
  - It is always a good idea to save your project documents in a dedicated file folder. So browse to a folder of your choice to save.
    Note: After you finish saving to the hard drive as described in these instructions, it is *also* a good idea to save the file to a flash drive and/or other location as well, ensuring that you have it saved in at least two separate locations.

1. Confirm you are still on the correct profile by clicking in the very top-left-hand corner on **Revu**, then hover over **Profiles** and confirm you are on the Real World Bluebeam Profile, indicated by the checkmark.

2. In a web browser, navigate to earth.google.com and search **860 Dolwick Drive, Erlanger, Kentucky, 41017**. Resize and reorient the view to grab the entire jobsite. Take a screenshot of the Google Earth image. Depending on when the images have updated, it should look something like the example shown here.

Note: If these steps for creating the page from Google Earth are not possible, skip ahead to the ***Alternate Option*** after step 9.

3. Open the Thumbnails Panel from the Panel Access Area and click on the Title Page.

4. Right click on the Title Page and select [Insert Blank Page...   Ctrl+Shift+N]

5. In the **Insert Blank Page** dialog box, we will insert a page the same size as the others in this drawing set. In the **Page** section enter the following:

6. In the bottom section, enter the following:

7. Next paste the screen shot previously captured from Google Earth, in the center of the blank page leaving a small white border around the image.

8. Right click on the image, hover on **Alignment** , and then select **⊕ Center in Document**

9. In order to keep the image where it is now placed, serving as a background for further markups, we must **Flatten** the image. Right click on the image and click **⚖ Flatten**

Alternate Option

If you are not able to create the page using an image taken from Google Earth as described in the preceding steps, one has been provided in the sample documents.

   i. Navigate to the sample documents provided into the **Lab 4** folder and open the document named **Lab 4 Site Logistics Insert Page**

   ii. Open the Thumbnails Panel from the Panel Access Area and right click on the page thumbnail, then click **Copy Pages**

   iii. Now go back to your Real World Bluebeam Revu – (your name) Lab #4 CD Drawings document and right click on the Title Sheet (first page) thumbnail and select **Paste Pages**

We will now begin to place markup objects on this sheet to show the planned location for site logistics items. Refer back to the Markups section in Lab 3 for instruction on adding Markups. A sample of the site logistics markups has been provided for reference following these steps.

10. Jobsite Office Trailers: Insert (2) rectangles in the center near the bottom of the sheet. Use the blue grip at one end of the rectangle to rotate it and align it as necessary (see right).

- Make the Line Color and Fill Color Yellow.
- Use a Callout to label these **Jobsite Office Trailers.**
- The Line Color and Text Color should be Blue.
- The Fill Color should be Yellow.
- The Line Width should be **3.00**.
- The Font Size should be **36**.

11. Crane Pads: Insert (2) squares, centered at the back side (opposite the road) of the two building masses. Align them to the back of the building.
- Make the Line Color and Fill Color Green.
- Use a Callout with (2) leaders to label these **40' x 40' Gravel Crane Pads for Steel Erection**.
- The Line Color and Text Color should be Blue.
- The Fill Color should be Green.
- The Line Width should be **3.00**.
- The Font Size should be **24**.

12. Laydown Area: Insert a square Text Box equivalent to the size of (4) of the crane pad squares. Place this square in the top-right-hand corner of the image near the edge of the paved area. In the box type **Laydown Area**.
- Make the Fill Color Magenta.
- The Line Color and Text Color should be Blue.
- The Line Width should be **3.00**.
- The Line Style should be Dashed 3.
- The Font Size should be **36**.

13. Deliveries Route: Place Arrows to show the route for deliveries to follow through the site.
    - The arrows should be **Blue** (Line Color and Fill Color).
    - The Line Width should be 20.00.
    - Make the End size 100%.
    - Use a Text Box with the word **Deliveries**, near the entrance to the site as a label.
    - The Line Color and Text Color should be **Blue**.
    - The Fill Color should be Pastel Cyan.

14. Contractor Parking: Using a Polygon, create a shape for the Contractor Parking. Right click on segments of the polygon and Convert to Arc to fine-tune the shape. Place a Text Box in the middle of the shape as a label.
    - For both the Polygon and the Text Box, the Line Color and Text Color should be **Blue**.
    - The Fill Color should be **Red Orange**.
    - The Line Width should be **3.00**.
    - The Font Size should be **36**.

15. Gravel Work Surface: Using the Highlight tool, create a line just outside the footprint of the building.
    - The Color should be **Blue**.
    - The Line Width should be **50.00**.
    - Confirm the Opacity is **100%** and the Highlight box is checked, as shown here.

    Use a Callout to label this **12' #2 Gravel Work Surface**.
    - The Line Color and Text Color should be **Blue**.
    - The Fill Color should be **Green**.
    - The Line Width should be **3.00**.
    - The Font Size should be **36**.

16. Dumpsters: Insert (3) rectangular Textboxes to represent dumpsters. Enter in them the word **Dumpster**. Place them spaced out along the edge of the pavement
    - The Line Color and Text Color should be **Blue**.
    - The Fill Color should be **Red**.
    - The Line Width should be **3.00**.
    - The Font Size should be **18**.

17. Temporary Facilities: Insert a rectangular shaped Cloud on the sidewalk in the front and the rear of the building.
   - The Line Color and Fill Color should be **Violet**.
   - Place a third matching cloud in the right margin and with it place a textbox labeling the shape: **Location of Temporary Bathroom Facilities**.

18. Jobsite Signage: Insert a rectangular shaped Cloud near the entrance drive to the site and to the office trailers.
   - The Line Color and Fill Color should be **Yellow**.
   - Place a third matching cloud in the right margin and with it place a textbox labeling the shape: **Location of Jobsite Identification Signage**.

19. Safety Barricades: Using a Polyline, trace the leading edge of the steep elevation change at the bottom-right-hand corner of the image where the detention basin is located.
   - The Line Color and Fill Color should be **Blue**.
   - The Line Style should be Fenceline Square
   - The Line Width should be **3.00**.

   Use a Callout to label this **Safety Barricade**.
   - The Line Color and Text Color should be **Blue**.
   - The Fill Color should be **Light Yellow Orange**.
   - The Line Width should be **3.00**.
   - The Font Size should be **24**.

Creating Spaces

Spaces can be created in order to organize the drawings into areas that can be referenced. Most commonly spaces are used to reference rooms or specific locations of various items.

1. If it is not already, open your file: Real World Bluebeam Revu – (your name) Lab #4 CD Drawings, and navigate once again to page 3, A111 FLOOR PLAN – FIRST FLOOR.

2. From the Panel Access Area, open the Spaces panel (⊞).

3. Click on the Add Space icon (⊞₊).

4. A crosshair curser will appear. Begin establishing the first space by clicking in each corner of room LOADING 151. Double click to conclude the first shape.

5. In the Add Space dialog box, enter a name for the space. Next to Space: enter **LOADING 151**.

Add Space	✕
?	Select Space Name
	Space: LOADING 151
	OK Cancel

6. Spaces can also be created by using the Dynamic Fill tool. Click on the Dynamic Fill icon (⊞) to open the tool. For a refresher on using the tool, refer to this item in Lab 3 – Tools.

7. We will now use the Dynamic Fill tool to create a space for room ELECTRICAL 143. Create a boundary (⊡) at the door to the room and then proceed to use the fill tool (⬙) to fill the room.

8. With the room filled, in the floating Dynamic Fill Toolbar, click the Spaces icon (⊞ ⌄) next to Create:, and then enter a name for the space as seen here.

9. Then click **Apply**

10. Continue to use either method to create a space for each room on the first floor.

Building Issue Tracking

Punchkeys

Now with Spaces established, those spaces can be referenced. One very useful way to reference the Spaces is for tracking issues. Issues are regularly tracked throughout the construction project. Issues will generally have a date they were created, a description, a responsible party to whom they are assigned, and often a due date. Similarly, at the end of a project, a punchlist is created. Punchkeys can be useful for any such tracking. To explore more about this functionality, we are going to look at the built-in Punchkeys that come with Bluebeam, and then learn how to create others.

1. If it is not already, open your file: Real World Bluebeam Revu – (your name) Lab #4 CD Drawings, and navigate once again to page 3, A111 FLOOR PLAN – FIRST FLOOR.

2. So far throughout this manual we have been working in the profile that we created. Now we will utilize one of the built-in profiles in Bluebeam. To change the profile, click in the very top-left-hand corner on **Revu**, then hover over **Profiles**. We will now click on **Field Issues**

3. Then, from the Panel Access Area, select the Tool Chest (💼).

Note that Bluebeam has built-in Tool Sets that include Punchkeys categorized into sets – Carpentry Issues, Electrical Issues, Flooring Issues, and so on. Click on one to see how it behaves.

4. Go ahead and click under Carpentry Issues on the (WR) and place it anywhere on the drawings.

5. Click to open the Properties Panel (⚙) and note some properties associated with the item. The subject is Carpentry which will categorize the item alongside others with like subjects. The comments field includes a description for the item.

Now we will create a new category and add some more punchkeys.

6. Once again select the Tool Chest (🗄). Then click on the drop down next to **Tool Chest ∨** and select **Manage Tool Sets**

7. In the Manage Tool Sets dialog box, click **Add...** . Next to Title: type **Concrete Issues**, click **OK** , and click **Save**

8. Click **Add...** again. This time type **Sprinkler Issues**, click **OK** , and click **Save**

9. Repeat the process one more time to add a tool set called **Wall Finish Issues**, and then click **OK** when finished.

Find where you had placed the (WR) previously and click on it. We will use this as a starting point for the new punchkeys we will now create.

10. With the (WR) selected, click the icon to open the Properties Panel (⚙).

11. Change the subject to **Concrete** and in the comments field enter **Concrete Crack**.

12. In the item, change the WR to **CC**.

13. Right click on the item, hover your curser over **Add to Tool Chest >**, and then select **Concrete Issues**

14. With the item still selected, now change the CC to **RE**. In the comments field enter **Rebar Exposed**.

15. Click on the icon to open the Properties Panel to confirm the subject is still Concrete.

16. Right click on the item, hover your curser over **Add to Tool Chest >**, and then select **Concrete Issues**

17. Return to the Tool Chest panel to confirm both punchkeys have been added to the Tool Set called Concrete Issues as shown here.

18. Next, with the RE item still selected, change the RE to **SE**.

19. Click the icon to open the Properties Panel (⚙). In the Subject: enter Sprinkler. In the comments type **Sprinkler Escutcheon Missing/Damaged.**

20. Right click on the (SE), hover your curser over Add to Tool Chest ＞, and then select Sprinkler Issues

21. Next, with the (SE) item still selected, change the SE to **DP**.

22. Click the icon to open the Properties Panel (⚙). In the Subject: enter Wall Finish. In the comments type **Drywall Patch.**

23. Right click on the item, hover your curser over Add to Tool Chest ＞, and then select Wall Finish Issues

24. Next, with the DP item still selected, change the DP to **TUP**.

25. Click the icon to open the Properties Panel (⚙). In the Subject: enter Wall Finish. In the comments type **Touchup Paint.**

26. Right click on the item, hover your curser over Add to Tool Chest ＞, and then select Wall Finish Issues

27. With the Tool Chest open and the new Tool Sets visible, perform a screenshot now to be submitted.

* Perform a screen shot and save to your computer.

Legends

A Legend can be used to create a snapshot table of all of the punchkeys on a page or on a set of pages.

We will now simulate a punchwalk that has been performed with punchlist items added as punchkeys to the drawings PDF file.

28. If it is not already, open your file: Real World Bluebeam Revu – (your name) Lab #4 CD Drawings, and navigate once again to page 3, A111 FLOOR PLAN – FIRST FLOOR.

29. Confirm you are still on the correct profile by clicking **Revu** in the very top-left-hand corner, hover over **Profiles**, and confirm **Field Issues** is checked.

30. Adding punchkey items on the drawings in Bluebeam is extremely simple. Click on the icon to open the **Tool Chest** (💼).

31. Scroll down to **Carpentry Issues** and click on the (DL) punch item.

32. Click to place the item near door (143) from Electrical room 143 to Loading room 151.

In reality, it is best practice to place the punchkey item as precisely as possible to best illustrate the real location of an issue in the field.

33. Click again to place the item near door (160B) from Training 1 room 161 to Lobby Storage/Table & Chairs room 160, and another near door (177) from Collaboration room 175 to Huddle room 177.

34. Next, in the Tool Chest, scroll down to **Electrical Issues** and click on the (EI) punch item.

35. Click to place the item near door (134) from Remit room 134 to Corridor room 130.

36. Next Scroll down to ⌄ Flooring Issues and click on the (TR) punch item.

37. Click to place the item in Men's Restroom room 121.

38. Next Scroll down to ⌄ Plumbing Issues and click on the (PI) punch item.

39. Click to place the item in Women's Restroom room 122.

40. Next Scroll down to ⌄ Concrete Issues and click on the (CC) punch item.

41. Click to place the item near the main stair in the center of Main Lobby room 111.

42. Next Scroll down to ⌄ Sprinkler Issues and click on the (SE) punch item.

43. Click to place the item near the center of Training 1 room 161.

44. Click again to place near the center of Training 2 room 162, Open Office room 180, and another in Huddle room 177.

45. Next Scroll down to ⌄ Wall Finish Issues and click on the (DP) punch item.

46. Click to place the item in Corrido room 130 and Restroom room 123.

47. Click again to place in DAV Museum room 116, Open Office room 180, and another in Seating room 171.

48. Next Scroll down to ⌄ Wall Finish Issues and click on the (TUP) punch item.

49. Click to place the item in Corrido room 170 and Restroom room 172.

50. Click again to place in Conference room 174, PO3 room 181, and another in Build room 186.

51. Zoom out so you can see the entire page and use the keyboard shortcut **Ctrl+A** to select all on the page.

52. While all items are still selected, right click on any of the items, hover over ⊞ Legend   >, and then click Create New Legend

53. Click to place the legend in the open area of the top-right-hand of the page.

54. With the legend still selected, it can be customized using the Properties Toolbar. Click the Title ∨ dropdown and next to Title:, type **Punchlist Legend**.

55. Click on the Columns ∨ dropdown, then select the dropdown next to Description:. Scroll down and click Comments

56. Click on the Columns ∨ dropdown and select Edit Columns

57. In the Edit Legend Columns dialog box, uncheck the checkbox next to Unit. Only Description and Quantity should be checked.

Create and Distribute Punchlist

With the punchlist items added as punchkeys in the drawings, and now with a legend created as a snapshot, the next step would be distributing the punchlist. This could be done as simply as extracting the pages where the items and legend are shown, and emailing as needed. However, it is often desirable to create a list of items that can be more easily distributed and reviewed.

58. If it is not already, open your file: Real World Bluebeam Revu – (your name) Lab #4 CD Drawings, and navigate once again to page 3, A111 FLOOR PLAN – FIRST FLOOR.

59. Confirm you are still on the correct profile by clicking Revu in the very top-left-hand corner, hover over Profiles , and confirm Field Issues is checked.

60. Click on the icon for the Markups List (⬚).

61. There can be many markups throughout a set of drawings. Click on Filter List, click the All above Page Label ⌄, and select our current page, **A111 FLOOR PLAN – FIRST FLOOR.**

62. As we've seen previously in the Markups List, we now want to configure columns. Click the dropdown next to Markups List ⌄, hover over Columns >, and then select Manage Columns...

63. In the Manage Columns dialog box, click the Custom Columns tab, and then click Add

64. Next to Name: , type **Resp. Subcontractor**.

65. Click the dropdown next to Type: and select Choice . This will create a dropdown selector in the newly created column.

You will now add each of the options that will appear in the dropdown selector. Entering each of them one at a time now will save you from having to type each of them for all punch items later.

66. Click Add

67. Next to Item: , enter the first subcontractor we will use for our punchlist: **Concrete Contractor**.

68. We can associate the dropdown option being created with specific data from the Subject column. By doing this, we can limit what items appear for each punchlist item in the dropdown selector we are creating. Recall that for our Punchkey data, the Subject column was used to establish categories of punchlist items. Therefore, next to Subject: , type **Concrete**.

69. Click OK , then OK , and then OK again.

70. In the Markups List you will now see a column for Resp. Subcontractor . In the row for the Concrete Crack item, you should be able to click in the Resp. Subcontractor column and select Concrete Contractor as the responsible subcontractor.

71. Once again, click the dropdown next to Markups List ∨ , hover over ⊞ Columns > , and then select Manage Columns...

72. In the Manage Columns dialog box, click the Custom Columns tab. This time click Resp. Subcontractor , and then click Modify

73. Click Add , and follow the steps to add each responsible subcontractor next to Item: , along with the appropriate subject associated with each as shown here.

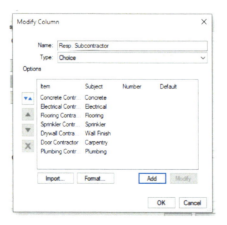

74. Click 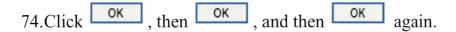 OK , then OK , and then OK again.

75. Proceed to click in the **Resp. Subcontractor** column and select the responsible subcontractor for each of the punchlist items.

Looking at the Markups List at this point, you can see how this provides all of the information needed to distribute as a punchlist for each contractor.

76. With the Markups List still open, click the Summary button (⬆) at the top of the Markups List area, and then select CSV Summary.

77. In the Markup Summary dialog box, choose the destination where you would like to save the export file in **Export to:**

78. In **File name:**, name the file **Real World Bluebeam Revu – (your name) Subcontractor Punchlist Export**.

79. Click the radio buttons to turn on Markups & Totals.

80. Leave all other defaults and then click OK

Stamps

The Stamp tool in Bluebeam can be exactly what it sounds like. Much like the old rubber stamps that can be customized and used to duplicate the same image/text simply and repeatedly, the Stamp tool in Bluebeam is a simple, yet effective tool.

1. We will now return to our custom profile by clicking in the very top-left-hand corner on **Revu**, then hover over **Profiles** and click on the **Real World Bluebeam Profile**.

2. If it is not already, open your file: Real World Bluebeam Revu – (your name) Lab #4 CD Drawings, and navigate once again to page 3, A111 FLOOR PLAN – FIRST FLOOR.

3. From the toolbar on the right, you'll find the Stamp icon (⊟∨) behaves as a dropdown menu when clicked. Click the Stamp tool and then select [Preliminary] Preliminary.pdf

4. Click to place the stamp in the bottom-right-hand corner of the sheet, just to the left of the sheet name and number.

Any stamp can be placed anywhere on a document in this manner. For our purposes, this particular stamp needs to be applied to all sheets.

5. Right click on the stamp and click [Apply to Pages...]

6. Leave All Pages in the Page Range and click [OK]

7. Scroll through all pages and reposition the stamp as necessary to be visible.

There are a number of simple stamps included in Bluebeam, but it is often useful to create a custom stamp.

8. From the toolbar on the right, click the Stamp tool (⊟∨) and then select [Create Stamp...]

9. In the Create Stamp dialog box, type **Submittal Stamp** next to [Subject:], change the width to **2** and the height to **2**. Then change the Font Color and the Line Color to **Blue**. Then click [OK]

10. Using the Markup tools, start by creating a rectangle just inside the perimeter of our 2"x2" space. The Line Color should be Blue. There should be no Fill ([▱]). The Line Width should be 1.00.

11. Insert a Text Box in the top two-thirds of the stamp. The Line Width will be **0.00** and therefore not visible. The Text Color should be **Blue**. Be sure the Autosize Text is on ([A]). Type the following for the text:

Reviewed for General Acceptance Only. This Reviewer does not relieve the Subcontractor of the responsibility for making the work conform to the requirements of the contract. The subcontractor is responsible for all dimensions, correct fabrication, and accurate fit with the work of other trades.

12. At the bottom of the stamp insert another Text Box. When you do so, this time notice that a selector appears Dynamic ▼ . Click the dropdown and select User . After that, hit the space bar twice and then click the dropdown again and select Date . Highlight the &[User] and &[Date] text and set the Font Size to 8.

Your stamp should look similar to that shown here

13. Click Save (💾)

          * Perform a screen shot and save to your computer.

Submittal Review

The overall submittal review process, will vary by Construction Management firm, including documentation of soliciting, receiving, reviewing, forwarding for design team review, and ultimately returning to the appropriate contractor and/or releasing for purchase. As a part of that process, Bluebeam can help by providing a solution for reviewing and tracking documents.

1. Navigate to the sample documents provided and open the file named **Lab 4 _ Roppe-700-Series-TPR-Wall-Base—-4-Cove—-Submittal.pdf**.

2. Perform a **File → Save As** and rename the file: **Lab 4 _ Roppe-700-Series-TPR-Wall-Base—-4-Cove—-Submittal_(your name)**.

This submittal document has been provided by the Flooring Contractor as documentation of the rubber base to be provided on the project. We can confirm this is the correct product based on the information from sheet A900 FINISH LEGEND AND ROOM FINISH SCHEDULE.

Now to process this submittal as reviewed, we will stamp it as such and pass it along to the design team for their review/approval.

3. From the toolbar on the right, click the Stamp tool (🔨˅) and then select ▢ Submittal Stamp.pdf

4. Click to place the stamp in an appropriate location near the bottom of the page.

5. Save this document to be submitted to your instructor.

RFI Posting

A Request for Information is used for many reasons on a construction project. One such reason might be due to a question that has come up onsite, possibly between a subcontractor working onsite and the Construction Management staff. It is vitally important that the question is documented, channeled to the appropriate party(s) for response, and ultimately fed to the proper folks onsite. In such an occasion, documenting the response in such a way that all necessary parties are adequately informed is as important as posing the question in the first place.

1. If it is not already, open your file: Real World Bluebeam Revu – (your name) Lab #4 CD Drawings, and navigate once again to page 3, A111 FLOOR PLAN – FIRST FLOOR.

2. From the sample documents, open the file named **Lab 4 RFI.pdf**.

3. Open the Thumbnails Panel from the Panel Access Area and right click on the page thumbnail, then click Copy Pages

4. Go back to your Real World Bluebeam Revu – (your name) Lab #4 CD Drawings document and right click on the page 3, A111 FLOOR PLAN – FIRST FLOOR. thumbnail and select Paste Pages . The RFI sheet should be inserted now into the page 4 position.

5. On page 3, A111 FLOOR PLAN – FIRST FLOOR, locate room ELECTRICAL 143. To the left of this room, insert a Text Box just outside the footprint of the building, and type **RFI 0015**.
   - Change the Line Color and the Font Color to Blue.
   - The Line Width should be **1.00**.
   - The Line Style should be **Cloud**.
   - The Line Size should be **1.00**.
   - The Font Size should be **12**.

6. Right click on the object and select Edit Action...   Ctrl+Shift+E

7. In the Action section, click the radio buttons for ⦿ Jump to and ⦿ Page . Next to Page, select **4**.

8. Make sure to save your work before moving on.

At this point you are finished with Lab 4. You will submit this document to your instructor along with all other Lab 4 documents.

Lab 5

Content Editing

Edit PDF Content

Bluebeam has powerful PDF Content Editing capabilities that make simple edits to PDF content fast and easy. Keep in mind that extensive modifications to PDFs is better done through the software in which they were originally created.

If it is not already open, follow the directions to open Bluebeam. From the sample documents provided, open the **Lab 5_Specifications.pdf**.

- Immediately perform a **File→Save As**, or simply select the command button from the toolbar.

- In the File name: field, enter **Real World Bluebeam Revu – (your name) Lab 5_Specifications.pdf**.

  - It is always a good idea to save your project documents in a dedicated file folder. So browse to a folder of your choice to save.
    Note: After you finish saving to the hard drive as described in these instructions, it is *also* a good idea to save the file to a flash drive and/or other location as well, ensuring that you have it saved in at least two separate locations.

1. Confirm you are still on the correct profile by clicking in the very top-left-hand corner on **Revu**, then hover over **Profiles** and confirm you are on the Real World Bluebeam Profile, indicated by the checkmark.

2. From the **Menu Bar**, click on **Edit**. Then hover over  and notice all of the options for editing PDF content (shown here).

3. In the document, in section **1.2 SUMMARY**, under A. Section Includes , notice that there is an error in the numbering, as number 5 follows number 3. From the **Menu Bar**, click on **Edit**. Then hover over PDF Content > and click on A|B Edit Text

4. Place your curser just after the 5, hit **Backspace** and then type **4**.

5. Next we would like to remove superfluous information from B. Related Sections . From the **Menu Bar**, click on **Edit**. Then hover over PDF Content > and click on Erase Content .

6. Using the crosshair selector, place a box around numbers 8 and 9 as shown here.

> 8. Divisions 23 - HVAC Air Distribution
> 9. Division 26 - Electrical

7. Next from the **Menu Bar**, click on **Edit**, and this time hover over Check Spelling > . Then click

8. Click OK and once the spell check has run, notice at the bottom of page one there is a typo substution

9. Once again, from the Menu Bar, click on Edit. Then hover over PDF Content > and click on AB Edit Text

10. Place your curser between the letters t and u, and insert the letters **it**.

11. Next, navigate to the second page. Near the top of the page, find a section called 1.3 REFERENCES and in it A. American Society for Testing and Materials (ASTM): . Notice that within the list of material test, the ASTM has been entered as ATSM. Rather than edit each of these individually as we have done previously, we will edit all of them at once.

12. From the Panel Access Area, select the icon for the Search panel ( ).

13. In the search field enter **ATSM** (the words we are trying to correct) and then press Enter.

14. Under ∨ Results you'll find each instance of this misspelled acronym. At the top of the list, check the **Select All** checkbox ( ) and then click the **Replace Checked** icon ( ).

15. In the empty field after Replace 'ATSM' With: , enter the correct acronym **ASTM**.

16. Click OK

17. In the Search panel, again run the search for ATSM, to confirm that all have been corrected.

18. Near the bottom of the third page, find D. Acoustical Certifications: . . . We will use a Strikethrough as this will not be required. Once again, from the Menu Bar, click on Edit. Then hover over PDF Content > and click on IA Select Text (this can also be found on the Navigation Bar).

19. Next, highlight all of the text in the paragraph after D. Acoustical Certifications:

20. Then right click on it and select **T̶ Strikethrough Selected Text**

21. As is commonly done in many documents, Bluebeam can add Headers and Footers to PDF documents. From the **Menu Bar**, click on **Document**. Then hover over **☐ Headers & Footers ＞** and click on **Add**

22. In the **Batch: Header and Footer** dialog box, click **Next** to apply to all pages in the document.

23. In the **Header and Footer** dialog box, in the **Header/Right** field, type **Real World Bluebeam Project**.

24. Next, place your curser in the **Footer/Left** field, and click **Date**

25. In the **Select Date Format** dialog box, select **MM/dd/yyyy** and then click **OK**

26. Then, place your curser in the **Footer/Center** field, and click **Page Number**

27. In the **Page Number Format** dialog box, select **1 of n** and then click **OK**

28. Make sure to save your work before moving on.

29. Save this document somewhere that you can easily find it to be submitted to your instructor

Sketching

Sketching is a functionality of Bluebeam that can have many applications. Here we will explore two basic applications that you can easily build onto to apply differently as needed.

1. If it is not already, follow the instructions to open Bluebeam.

2. From the Menu Bar, click on File. Then click 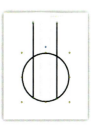 New PDF...

3. In the New dialog box, click

4. From the Markup tools, use a circle and two lines to create a receptible symbol as shown here. For each shape, the Line Color should be **Black**, and the Line Width should be **1.00**.

5. From the Navigation bar, click the Select tool ( ). Then click and drag a rectangle around your symbol to select all 3 shapes making up the symbol.

6. Right click on this selection and click Group   Ctrl+G . Notice that the individual shapes have now become one combined symbol.

7. Next, right click on the newly created symbol and click Change Colors

8. In the Change Multiple Colors dialog box, next to Destination Color: select **Blue**. Then click **OK**

9. Right click on the shape again, hover over Add to Tool Chest > and then select Construction Symbols

10. Click on the icon for the Tool Chest ( ) to confirm your symbol has been added to the Construction Symbols Tool Set as shown here.

11. With the Tool Chest still open, perform a screenshot

* Perform a screen shot and save to your computer.

Sketches can be as simple as this symbol or as complicated as a Structural Steel Connection Detail. It is often necessary to complete a sketch that is to scale.

12. We will start this next exercise by accessing the drawings used in Lab 3. Navigate to the sample documents and open the file from the Lab 3 folder called Lab 3 CD Drawings.pdf.

13. Navigate to the second page, A111 FLOOR PLAN – FIRST FLOOR, and find Conference room 174. We will use this room as the basis for our scaled sketching.

For this exercise, we need to confirm that a conference table will fit properly in this conference room.

14. From the Menu Bar, click on File. Then click New PDF...

15. In the New dialog box, click OK

16. From the Menu Bar, click on File and Save-As. Name your file **Real World Bluebeam Revu – (your name) Lab 5_Conference Room Sketch.pdf**.

17. In the Lab 3 CD Drawings document, zoom to place the Conference room 174 in the center of your workspace.

18. Then from the Menu Bar, click on Edit. Then hover over  PDF Content and click on  Snapshot Content   Shift+G

19. Using the crosshair selector, place a box around the room as shown here.

Take your time to capture all of the conference room itself without too much ecess space outside of the room. It may take a few tries. Note the rectangle in blue.

20. Back to your newly created document, right click and select ⬚ Paste   Ctrl+V

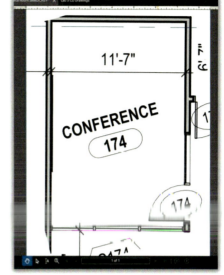

21. Click the yellow grip on the corners to resize the room to nearly fill the page. Then click the blue grip to rotate the image.
    Note: You will find it takes some trial and error. But if you hold the shift key while rotating, you can fine-tune the alignment. It should look similar to that shown here.

22. Right click on the object and then click ⬚ Flatten

23. In order to create a sketch that is to scale, we must first establish a scale for the sheet. From the Panel Access Area, click the icon to open the Measurements Panel (▭).

24. Click ⬚ Calibrate

25. Then ⬚ OK

26. We can use the dimension across the width of the room to establish the scale. Click at each end of the dimension line. Then when prompted, enter **11'-7"**

27. Click ⬚ Apply Scale

Now that we have the room created to scale, we would like to confirm that the selected conference table will fit in the center of the room with at least 42" of clearance all the way around.

To establish the 42" clearance around the room, we will use rectangle shapes to show the clearance.

28. From the Menu Bar, click on Tools. Then hover over ⬚ Sketch to Scale   ＞ and click ⬚ Rectangle Sketch to Scale

29. Click in the top-left-hand corner of the room. A dialog box will appear like that shown to the right. Under Width, enter **0'-42"**. Under Height, enter **17'-0"**, and then hit **Enter**.

30. On the Panel Access Area, click the icon to open the Properties Panel (⚙).

31. Change the Fill Color to **Blue** and the Fill Opacity to **10%**. Change the Line Width to **0.00**.

32. At the bottom of the panel, click [Set as Default]

33. Follow these same steps to create another scaled rectangle and position it against the right side of the room.

34. Next once again, from the Menu Bar, click on Tools. Then hover over [Sketch to Scale >] and click [▢ Rectangle Sketch to Scale]

35. Click in the top-left-hand corner of the room. A dialog box will appear like that shown to the right. Under Width, enter **11'-6"** Under Height, enter **0'-42"**, and then hit **Enter**.

36. Do the same to create another scaled rectangle and position it at the bottom of the room.

37. From the sample documents, open the file called **Lab 5 Conference Table Selection.pdf**.

In the PDF cut sheet, note that the selections have been notated by blue clouds. We see that the table is 96 x 48".

38. Next once again, from the Menu Bar, click on Tools. Then hover over [Sketch to Scale >] and click [▢ Rectangle Sketch to Scale]

39. Click anywhere this time. A dialog box will appear like that shown to the right. Under Width, enter **0'-48"** Under Height, enter **0'-96"**, and the hit **Enter**.

40. On the Panel Access Area, click the icon to open the Properties Panel (⚙).

41. Change the Fill Color to **Green** and the Fill Opacity to **10%**. Change the Line Width to **0.00**.

42. Reposition the green shape in the center of the room to confirm that it will fit within the area left with the established clearance.

43. Make sure to save your work before moving on.

Forms

Forms is a powerful functionality of Bluebeam that can be used to create electronic forms or convert legacy forms into a form that can be completed electronically.

Previously you saw an RFI form used in Lab 4 for posting an RFI in the drawings. Let's take a look at how we can customize this PDF form to be completed as an electronic, fillable form in Bluebeam.

1. If it is not already open, follow the directions to open Bluebeam. From the sample documents provided, open the **Lab 5_RFI-Template.pdf**.

2. Immediately perform a **File→Save As**, or simply select the command button from the toolbar. In the File name: field, enter **Real World Bluebeam Revu – (your name) Lab 5_RFI Form.pdf**

3. Confirm you are still on the correct profile by clicking in the very top-left-hand corner on **Revu**, then hover over **Profiles** and confirm you are on the Real World Bluebeam Profile, indicated by the checkmark.

4. From the Panel Access Area, click the icon to open the Forms Panel ().

Take a moment to get to know the commands located in this panel. Each command will create a different type of form field and has a unique function and set of properties.

- Text Box ▣ - The most basic and most commonly used form field, the Text Box offers an open field in which text can be entered. It can be sized as needed. Properties that can be selected include allowing Multi-line text entry, allowing Rich Text Formatting, limiting character entry, spell check, and others.

- Radio Button ◉ - A Radio Button is used as a simple toggle selection. It can be used to indicate a selection. With multiple Radio Buttons working together, they can be used to indicate a choice when only one option can be selected from a set of choices.

- Check Box ☑ - While similar to the Radio Button, when using a Check Box in a series of options, multiple Check Boxes can be selected.

- List Box ▣ - As its name indicates, a List Box is a box containing a preset list of options. The box can be sized and positioned as needed. When space requires it, the list can contain a scroll bar. The list can be set to allow a single selection only or Multiple Selection.

- Dropdown ▣ - Similar to a List Box, the Dropdown provides a preset list of options. A Dropdown will save in space as compared to a List Box. Another difference would be that a Dropdown would limit the user to selected only one option from the list. Allowing custom text entry is also an option.

- Button ▣ - Using this option, a button can be created with a customized look and the action performed by the button can be configured in the Properties Panel.

- Digital Signature ✗✎ - A powerful electronic signature option for documents requiring a level of security.

5. We'll start by installing the Text Boxes needed at the top of the form. From the Forms Panel (▣), click the icon to select Text Box (⊞), and place it on the line next to RFI No.: as seen here.

6. From the Panel Access Area, click the icon to select the Properties Panel (⚙). In the Name: field, enter **RFI No.**. Enter the same thing in the Tooltip: field as well.

7. From the Forms Panel (▣), click the icon to select Text Box (⊞), and place it on the line next to Date:

8. From the Panel Access Area, click the icon to select the Properties Panel (⚙). In the Name: field, enter **Date**. Enter the same thing in the Tooltip: field as well.

9. In the ⌄ Format section, click the dropdown next to Category and select **Date** and then the **mm/dd/yyyy** format.

The form seems unclear what is meant by Answer by:. To clarify this, we are going to replace this with Due Date.

10. From the Menu Bar, click on Edit. Then hover over PDF Content > and click on A|B Edit Text

11. Place your curser after the Y of Answer By. Hit the Backspace key to remove this text and then type **Due Date**.

12. Follow the previous steps to place a Text Box for Due Date, and for each address line after the To: and From:. Then update the Name: and Tooltip: fields for each.

13. Once again, in the ⌄ Format section, click the dropdown next to Category and select **Date** and then the **mm/dd/yyyy** format.

14. Go ahead and follow the previous steps to place a Text Box for Project, Ref. Drawing, and Ref. Spec.. Then update the Name: and Tooltip: fields for each.

15. From the Forms Panel (🖉), click the icon to select Dropdown (🔽), and place it on the line next to ATTN:

16. From the Panel Access Area, click the icon to select the Properties Panel (⚙). In the `Name:` field, enter **ATTN:**. Enter the same thing in the `Tooltip:` field as well.

17. In the Properties Panel, in the `∨ Options` section, we will enter the selectable options for this field. In the Items: field, first enter **Architect** and click `Add`

18. Follow these steps to also add **Structural Engineer**, **Civil Engineer**, **HVAC Designer**, **Electrical Designer**, and **Plumbing Designer**.

19. Then check the box for `☐ Allow user to enter custom text`

20. In the space under the From: fields, we will enter some additional fields. Using the Typewriter tool, enter the following: **Cost Impact:**, **Schedule Impact:**, **Draft: Yes   No**.

21. Right click on each and select `🔔 Flatten`

22. It should look similar to that shown here.

23. To the left of Cost Impact and Schedule Impact, we will place a Check Box. From the Forms Panel (🖉), click the icon to select Check Box (☑), and place it to the left of Cost Impact:.

24. Repeat these steps to place another Check Box to the left of Schedule Impact:.

25. In the Properties Panel, update the `Name:` and `Tooltip:` as done previously.

26. From the Forms Panel (▤), click the icon to select Radio Button (⦿), and place one to the left of each Yes and No.

27. In the Properties Panel, update the Name: and Tooltip: as done previously. For the **Yes** Radio Button, enter **Draft Yes**, and likewise for the **No** Radio Button.

28. Next we'll place a large Text Box in the space beneath both RFI Description and RFI Response. Follow the previous steps to place a Text Box each. Then update the Name: and Tooltip: fields for each.

29. From the Panel Access Area, click the icon to select the Properties Panel (⚙). In the ⌄ Options section, check the box for ☐ Multi-line

30. Next, from the Forms Panel (▤), click the icon to select Button (▭), and place near the bottom-right-hand corner of the sheet.

31. In the Properties Panel, update the Name: and Tooltip: as done previously. For each enter the word **Submit**.

32. In the ⌄ Appearance section, make the Fill Color Blue. The Font Color should be **20% Gray**.

33. In the ⌄ Options section, next to Label:, type Submit

34. In the ⌄ Actions section, click the Add action button.

35. In the Action dialog box, click on the **Form** tab. Click the radio button to select ○ Submit Form and then click Options

36. When the Submit Form Options dialog box appears, next to URL: type **mailto:** followed by your email address, with no spaces.

37. Leave all other defaults and click OK, then OK

38. The last item to add to this form is a signature line. Next, from the Forms Panel (▤), click the icon to select Digital Signature (✎), and place near the bottom-left-hand corner of the sheet.

39. In the Properties Panel, update the Name: and Tooltip: as done previously.

40. Make sure to save your work before moving on.

Let's take a look at the power of Forms tool in automatically creating form fields.

41. Navigate this time to the sample documents, in the Lab 5 folder and open the file called **Lab 5 _ JHA-Template.pdf**.

42. Immediately perform a **File→Save As**, or simply select the command button from the toolbar. In the File name: field, enter **Real World Bluebeam Revu – (your name) Lab 5_JHA-Template.pdf**.

43. From the Menu bar, select the Tools menu, hover over Form >, and then select Automatically Create Form Fields. The auto-creation should happen immediately.

44. Click to review that all Text Boxes and other form fields have automatically been created. Further customization could be done from here if desired.

45. Make sure to save your work before moving on.

Sending & Exporting

There are many useful ways to send and export PDFs along with various other file types in Bluebeam. We will look at some of the most common applications.

Sending PDFs from Bluebeam can be very simple. Much like other programs, an open document can be sent by simply clicking File, and then Email. For a recurring email, Bluebeam also has a feature called Email Templates.

1. We will do this by using the RFI form created previously. Navigate to open your Real World Bluebeam Revu – (your name) Lab 5_RFI Form.pdf document.

2. From the Menu Bar, select the File menu hover over Email Templates > and then click New Template...

3. In the Email Template dialog box, next to Template: enter **RFI Form**. Next to To: enter your email address. Next to Subject: enter **Bluebeam Project RFI**. In the Message: enter **Please see attached Project RFI Form**.

* Perform a screen shot and save to your computer.

4. Click OK

Another common application for exporting from Bluebeam would be exporting a PDF to another application, the most common being Word, Excel, or even PowerPoint.

We will try this out using an equipment schedule.

5. Navigate this time to the sample documents, in the Lab 5 folder and open the file called **M401_Mechanical Equipment Schedule.pdf**.

6. From the Menu Bar, select the File menu click Export >, then Excel Workbook >, and then Page Region

7. With the crosshair selector, click and drag to select the AIR DEVICE SCHEDULE as shown here.

Tip: Use your mouse wheel to zoom and pan to make the selection easier.

8. A dialog box will appear in order to save the export. In the File name: field, enter **Real World Bluebeam Revu – (your name) Lab 5_Export**.

9. Make sure to save your work before moving on. Submit this file, along with all other Lab 5 files, to your instructor.

Capstone

The *Real World Bluebeam for Construction Managers* lab manual provides the framework for a comprehensive Capstone Project. The Capstone Project gives the student a structure through which the student can demonstrate proficiency in the various skills learned throughout this manual. This section will not provide the detailed steps to perform each task, as was given in the previous labs, rather it will give only instructions on what tasks are to be performed.

Your instructor will provide the documents to be used throughout the Capstone Project. Your instructor will also provide instructions on how each step is to be submitted.

Document Management

1. From the documents provided, compile all drawing sheets into one PDF file.

2. From the compiled drawing set, create Page Labels for each page.

3. Once Page Labels have been created, these can be used to separate the overall drawing set into separate sets that will be smaller, more manageable files. Extract pages to create individual files based on design discipline.

4. From the overall floor plans, create Hyperlinks to the enlarged plans.

Markups

1. Create a Tool Set called Markup Tools containing the following tools:

Text Box
- The Line Color and Text Color should be **Blue**
- Change the Fill to **Magenta**
- Change the Fill Opacity to **30%**
- The Line Width should be **3.00**
- The Line Style should be **Dashed 3**
- The Font Size should be **36**

Callout
- The Line Color and Text Color should be **Blue**
- The Fill Color should be **Violet**
- The Line Width should be **3.00**
- The Font Size should be **18**

Cloud+
- Make the Fill Color **Pastel Cyan**
- Change the Fill Opacity to **20%**
- The Line Color and Text Color should be **Blue**
- The Line Width should be **3.00**
- The Line Size should be **1.00**
- The Line Style should be **Dashed 3**
- The Font Size should be **24**

Rectangle
- Make the Line Color **Red Orange**
- Make the Fill Color **Yellow**
- The Line Style should be **Dashed 3**
- Change the Fill Opacity to **10%**

Arrow
- The Line Color and Fill Color should be **Green**
- The Line Width should be **20.00**
- Make the End size **100%**.

2. Add Markups to the Floor Plan for the following design clarifications.

   Cloud+
   - Place cloud around a wall outside corner and in the callout enter **Need corner protection requirements**.

   Text Box
   - Place a Text Box at the mechanical/electrical/utility room and in the text **Need heating requirements**.

   Rectangle
   - Place a Rectangle around the stair column.

3. On the Floor Plan, set the scale for sheet and complete (3) Length measurements to confirm dimensions labeled on the sheet.

Field Use

1. Using the Site Plan, create a Site Logistics Plan including the following at a minimum.
   - Label Site Extents
   - Location for Dumpster(s)
   - Location for Temporary Bathrooms
   - Location for Site Entry Signage
   - Location for Laydown Area
   - Route for Deliveries
   - Location for Crane/Cranepad

2. Create a Markup object on the Floor Plan that hyperlinks to an RFI response.

Content Editing

1. Create a scaled sketch. Using
   - Using scaled Markup objects create a plan sketch of a ramp in plan view. The ramp should contain (1) switchback, clear a vertical rise of 3'-2" and have a path of travel with a width of no less than 48".

2. Create an electronic Transmittal Page containing fillable form fields. The form should contain the following at a minimum.
   - Text Boxes
     - Project Name
     - Project Number
     - Transmittal number
     - Date
     - To: name and address
     - From: name and address
     - CC:
     - Subject line
     - Message
     - Attachments:
   - Dropdown field for "Action:" field containing the following options.
     - For approval
     - For review
     - Submitted
     - For your use
     - As requested
     - Review and comment
   - Radio Button for Draft: Yes No
   - A Digital Signature line
   - A Button labeled "Submit" that will be emailed to yourself
   - A Header with "Real World Bluebeam Project" in the right position

Takeoff Labs

Using the drawings provided, you will perform a quantity takeoff of the following categories. Your instructor should provide a list of items to takeoff within each category.

1. Concrete
2. Masonry
3. Steel
4. Doors & Windows
5. Metal Studs & Drywall
6. Flooring
7. HVAC
8. Plumbing
9. Electric